I0055643

Fractional Dynamics
in Comb-like
Structures

Fractional Dynamics in Comb-like Structures

Alexander Iomin

Technion-Israel Institute of Technology, Israel

Vicenç Méndez

Universitat Autònoma de Barcelona, Spain

Werner Horsthemke

Southern Methodist University, USA

World Scientific

NEW JERSEY · LONDON · SINGAPORE · BEIJING · SHANGHAI · HONG KONG · TAIPEI · CHENNAI · TOKYO

Published by

World Scientific Publishing Co. Pte. Ltd.

5 Toh Tuck Link, Singapore 596224

USA office: 27 Warren Street, Suite 401-402, Hackensack, NJ 07601

UK office: 57 Shelton Street, Covent Garden, London WC2H 9HE

British Library Cataloguing-in-Publication Data
A catalogue record for this book is available from the British Library.

FRACTIONAL DYNAMICS IN COMB-LIKE STRUCTURES

Copyright © 2018 by World Scientific Publishing Co. Pte. Ltd.

All rights reserved. This book, or parts thereof, may not be reproduced in any form or by any means, electronic or mechanical, including photocopying, recording or any information storage and retrieval system now known or to be invented, without written permission from the publisher.

For photocopying of material in this volume, please pay a copying fee through the Copyright Clearance Center, Inc., 222 Rosewood Drive, Danvers, MA 01923, USA. In this case permission to photocopy is not required from the publisher.

ISBN 978-981-3273-43-6

For any available supplementary material, please visit
https://www.worldscientific.com/worldscibooks/10.1142/11076#t=suppl

Desk Editor: Christopher Teo

Typeset by Stallion Press
Email: enquiries@stallionpress.com

To Our Families

Preface

Du mouvement avant toute chose![1]

Osip Mandelstam

Movement is one of the most common features of nature, if not to say a way of existence, and random walks play an important role in this context. The mathematical theory of various types of random walks is well developed. Random walks often provide the underlying mesoscopic mechanism for transport phenomena in physics, chemistry and biology. In particular, anomalous transport in branched structures has attracted considerable attention. Combs are simple caricatures of various types of natural branched structures that belong to the category of loopless graphs. The comb model was introduced to understand anomalous transport in percolation clusters. Comb-like models have been widely adopted to describe kinetic processes in various experimental applications in medical physics and biophysics, chemistry of polymers, semiconductors, and many other interdisciplinary applications. We present a random walk description of the transport in specific comb geometries, ranging from simple random walks on comb structures, which provide a geometrical explanation of anomalous diffusion, to more complex types of random walks, such as non-Markovian continuous-time random walks. The simplicity of comb models allows us to perform a rigorous analysis and to obtain exact analytical results for various types of random walks and reaction-transport processes.

Due to their importance, random walks on combs have been studied for over three decades. The results are scattered over many publications, and it is difficult to keep abreast of the literature. The first goal of this monograph is to provide a synthesis of the field and a concise, but comprehensive,

---

[1] "Movement first of all" Mandelstam's paraphrasing of P. Verlaine.

description of its actual state. The second one is to present a contemporary and thorough review of random walks on comb-like structures, reflecting our current knowledge and describing the state-of-the-art of the research on comb models.

Another goal of the book is to present elements, methods, and techniques of random processes and their application to random walks in the framework of comb models, which can be solved exactly. These examples can also be considered as tutorial material. In this sense, the comb model is a mathematical realization of one of Occam's razor principles, expressed by Einstein: "Everything should be kept as simple as possible, but no simpler".[2]

The book is divided into four parts. The first part consists of three chapters providing a brief introduction to the field of anomalous transport and fractional calculus. This concise introduction facilitates understanding of the next chapters of the book and makes the presentation self contained. Chapter 1 contains a short history of the comb model, which is introduced as a simplified description of a percolation cluster. We also establish the relation of the comb model to continuous time random walk theory and fractional calculus. Chapter 2 provides a concise introduction to fractional calculus and special functions of fractional kinetics. Chapter 3 deals with Markov processes, which are one of the most important mechanisms of anomalous transport in the comb geometry.

The second part is devoted to various realizations of anomalous transport in the comb, such as subdiffusion, superdiffusion, ultra-slow diffusion, and stochastic localization. We use different methods of description of these anomalous kinetics, and throughout our treatment we present microscopic, mesoscopic, and macroscopic approaches to transport processes. Chapter 4 deals with a generic method of the mesoscopic description of random walks on combs. An important part of the analysis is to consider the generalized dynamics in the Laplace domain. Chapter 5 is devoted to the Langevin formalism, and we discuss how transport in ramified structures, including combs, can be described in the framework of Langevin equations. We describe a functional method to obtain transport characteristics, or coefficients, for the corresponding Fokker–Planck equation. In Chapter 6, we study the role of inhomogeneous advection, which gives rise to superdiffusion in a subdiffusive medium as well as localization of diffusive spreading. Ultra-slow diffusion and its destruction due to two-dimensional branch sub-

---

[2]Quoted from the book by E. S. Swanson, "Science and Society" [1].

diffusion is considered in Chapter 7.

The third part is devoted to transport with reactions. It contains two chapters where we deal with two examples of fractional reaction-diffusion equations. We describe realistic scenarios of reaction front propagation in the framework of a Fisher–Kolmogorov–Petrovskii–Piskunov scheme describing ionic transport in dendritic spines in Chapter 8 and actin polymerization in a comb microfluidic device in Chapter 9. The theoretical treatment of the experimental situations is based on the analytical method of hyperbolic scaling and the Hamiltonian–Jacobi technique for resolving reaction front propagation.

The final part consists of three chapters and is devoted to some extensions of the comb model. Specifically, we consider two directions. The first one in Chapter 10 is a geometrical generalization by means of additional temporal and spatial kernels in the framework of a generalized Fokker–Planck equation. The second extension concerns the framework of quantum mechanics on the comb. This leads to the time fractional Schrödinger equation studied in Chapters 11 and 12. In this way, we eventually arrive at the path integral description of both fractional diffusion and time fractional quantum mechanics.

This book is a result of our collaboration with many researchers, including extended discussions and joint publications. It is our great pleasure to thank our collaborators and colleagues, who shared with us their time, ideas, and knowledge and from whom we have learned so much. We especially acknowledge fruitful collaborative work and helpful and instructive discussions with Prof. Eli Barkai, Prof. Emmanuel Baskin, Prof. Daniel Campos, Prof. Sergei Fedotov, Prof. Shmuel Fishman, Prof. Holger Kantz, Prof. Ralf Metzler, Dr. Alexander Milovanov, Prof. Thomas Pfohl, Dr. Trifce Sandev, Prof. Igor Sokolov, and Dr. Vasily Zaburdaev. We also acknowledge financial support by the Israel KAMEA Program and the Israel Science Foundation (ISF).

Haifa, Israel   Alexander Iomin,
Barcelona, Spain   Vicenç Méndez,
Dallas, TX   Werner Horsthemke
                                                    January, 2018

Contents

PART 1

Introduction to Methods and Techniques

Chapter 1

A brief history and overview of methods

"We consider it a good principle to explain the phenomena by the simplest hypothesis possible."

C. Ptolemy

1.1 Introduction

Random walks often provide the underlying mesoscopic mechanism for transport phenomena in physics, chemistry and biology. A wide class of random walks gives rise to normal diffusion, where the mean-square displacement (MSD), $\langle (\Delta r)^2(t) \rangle$, grows linearly with time t for large times. In many important applications, however, the MSD behaves like $\langle (\Delta r)^2(t) \rangle \sim t^\alpha$, with $\alpha \neq 1$, and the diffusion is anomalous [2, 3]. Anomalous diffusion can be modeled by various classes of random walks [4]. We focus on the important class of continuous time random walks (CTRWs), which were introduced by Montroll and Weiss in 1965 [5] and since then have been discussed in many reviews for a large variety of applications, see for example Refs. [2, 3]. A specific feature of a CTRW is that a walker waits for a random time τ between any two successive jumps. These waiting times are random independent identically distributed (iid) variables with a probability distribution function (pdf) $\psi(\tau)$, and the tail of the pdf determines if the transport is diffusive ($\alpha = 1$) or subdiffusive ($\alpha < 1$). Heavy-tailed (in the form of a power law) waiting time pdfs give rise to subdiffusion. Realistic models of the waiting time pdf have been formulated for transport in disordered materials with fractal and ramified architecture, such as porous discrete media [6] and comb and dendritic polymers [7–9], and for transport in crowded environments [10].

A simple caricature of various types of natural branched structures that belong to the category of loopless graphs is a comb model (see Fig. 1.1). The comb model was introduced to understand anomalous transport in percolation clusters [11–15]. Now, comb-like models are widely employed to describe various experimental applications. These models have proven useful to describe the transport along spiny dendrites [16, 17], percolation clusters with dangling bonds [13], diffusion of drugs in the circulatory system [18], energy transfer in comb polymers [7,8] and dendritic polymers [9], diffusion in porous materials [19–21], the influence of vegetation architecture on the diffusion of insects on plant surfaces [22], and many other interdisciplinary applications. Random walks on comb structures provide a geometrical explanation of anomalous diffusion.

Diffusion on comb structures has also been studied by macroscopic approaches, based on Fokker–Planck equations [15], which have been applied to describe diffusive properties in discrete systems, such as porous discrete media [6], infiltration of diffusing particles from one material into another [23], and superdiffusion due to the presence of inhomogeneous convection flow [24–28]. Other macroscopic descriptions, based on renormalizing the waiting time pdf for jumps along the backbone to take into account the transport along the branches [29], have been found useful to model continuous-time-reaction-transport processes [30] and human migrations along river networks [31]. For example, a numerical study of the encounter problem of two walkers in branched structures shows that the topological heterogeneity of the structure can play an important role [32].

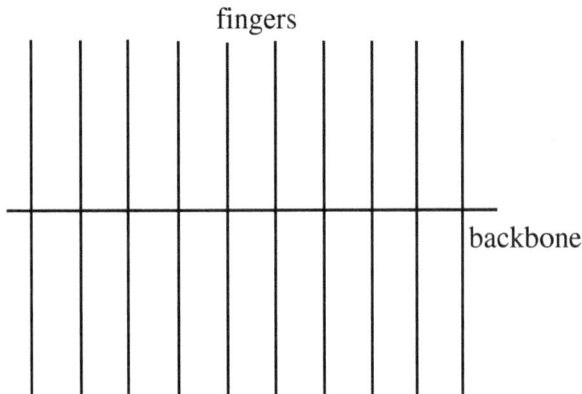

Fig. 1.1   Comb structure consisting of a backbone displayed in the x-direction and continuously distributed branches in the y-direction, which extend to infinity.

Another example is the occupation time statistics for random walkers on combs where the branches can be regarded as independent complex structures, namely fractal or other ramified branches [33]. Finally, we want to mention studies to understand the diffusion mechanism along a variety of branched systems where scaling arguments, verified by numerical simulations, have been able to predict how the MSD grows with time [34].

The comb model was introduced by Ziman [11] and generalized by White and Barma [12], Gefen and Goldhirsch [14], and Weiss and Havlin [13]. The model captures the main behavior of a random walker on a fractal medium, like a percolation cluster, where a random walker can move only onto conducting sites and is not allowed to step on non-conducting or isolating sites. At the percolation threshold, a percolation structure can be idealized as a single infinite cluster, consisting of a conducting path, which corresponds to a backbone, and side branches, or fingers with dangling bonds. At some level of idealization,[1] this structure corresponds to a comb as shown in Fig. 1.1.

Following this simplification process, a linear ferromagnetic chain with dead-end branchings of finite length L was originally suggested by Ziman in the context of spin wave propagation on percolation networks [11]. This comb model with equal length of the fingers was generalized in Refs. [12,14] to combs with a random finger length.[2] Weiss and Havlin [13] studied the transport properties of a discrete in space and time random walk on the comb as a simple model of anomalous diffusion on fractal structures. Considering simple random walks inside the fingers, they constructed a distribution of recurrences to the backbone, which eventually yields the waiting time pdf in the power law form $\psi(t) \sim 1/t^{3/2}$ for the CTRW on the backbone only. The interplay of the transport in the backbone and the fingers results in a memory effect in the backbone transport, which gives rise to subdiffusion with the MSD $\langle x^2(t) \rangle \sim \sqrt{t}$. We will discuss this comb mechanism for the waiting time pdf, based on discrete Markov chains, in greater detail in Chapter 3.

---

[1]There are two components in this simplification: (i) the system is loopless without connections between the fingers, and (ii) the fingers are uniformly spaced along the backbone.

[2]The most interesting case is the power law distribution $L^{-1-\gamma}$ of the finger lengths with $0 < \gamma < 1$, see, e.g., Refs. [35,36].

1.2   Subdiffusion vs diffusion

Arkhincheev and Baskin   [15] adopted a macroscopic phenomenological approach to describe comb-like structures. They introduced a standard two dimensional (2D) diffusion equation, which describes Brownian motion in the specific comb geometry shown in Fig. 1.1. A matrix of diffusion coefficients reflects the comb geometry, which implies that displacement in the x-direction is possible only along the structure axis, i.e., the x-axis at $y = 0$. Diffusion in the x-direction is highly inhomogeneous, and the diffusion coefficient is given by $D_{xx} = D_x \delta(y)$, while the diffusion coefficient in the y-direction, the side-branch direction, also called teeth or fingers, is a constant, namely $D_{yy} = D_y$. The diffusion equation on the comb structure reads

$$\frac{\partial}{\partial t} P(x, y, t) = D_x \delta(y) \frac{\partial^2}{\partial x^2} P(x, y, t) + D_y \frac{\partial^2}{\partial y^2} P(x, y, t), \qquad (1.1)$$

where $P(x, y, t)$ is the pdf of finding a diffusing particle at time t at the position with coordinates (x, y) in the 2D comb space. The initial condition is

$$P(x, y, t = 0) = \delta(x)\delta(y), \qquad (1.2)$$

and natural boundary conditions are chosen at infinity, i.e.,

$$P(x, y, t) = \partial_x P(x, y, t) = \partial_y P(x, y, t) = 0 \qquad (1.3)$$

for either $x = \pm\infty$ or $y = \pm\infty$.

Particular solutions of the equation can be obtained. Following standard procedures, we perform the Fourier transformation in space

$$\mathcal{F}_x\big[P(x, y, t)\big] = \tilde{P}(k, y, t) = \int_{-\infty}^{\infty} e^{ikx} P(x, y, t) dx \qquad (1.4)$$

and the Laplace transformation in time

$$\mathcal{L}\big[P(x, y, t)\big] = \hat{P}(x, y, s) = \int_{0}^{\infty} e^{-st} P(x, y, t) dt. \qquad (1.5)$$

In Fourier–Laplace space, Eq. (1.1) reads

$$s\hat{\tilde{P}}(k, y, s) = -D_x \delta(y) k^2 \hat{\tilde{P}}(k, y, s) + D_y \frac{\partial^2}{\partial y^2} \hat{\tilde{P}}(k, y, s) + \delta(y), \qquad (1.6)$$

where throughout the book, s is the Laplace variable and k the Fourier variable. We look for solutions of the form

$$\hat{\tilde{P}}(k, y, s) = f(k, s)g(y, s) = f(k, s) \exp\left[-\sqrt{s/D_y}|y|\right], \qquad (1.7)$$

where $f(k,s) = \hat{\tilde{P}}(k, y = 0, s)$ describes diffusion along the backbone in Laplace space. Substituting solution (1.7) into Eq. (1.6) and integrating with respect to y, we obtain an equation for $f(k,s)$,

$$f(k,s) = \frac{1}{2\sqrt{sD_y} + D_x k^2}. \tag{1.8}$$

Analytical solutions of the distribution functions can be represented in the form of the Mittag–Leffler and the Fox H-functions. This will be considered in detail in the next chapters.

Here we show that the transport along the backbone is subdiffusive, while normal diffusion occurs along the side-branches or fingers. To this end we calculate the MSDs,

$$\langle x^2(t) \rangle = \int x^2 P(x, y, t) dx dy, \tag{1.9a}$$

$$\langle y^2(t) \rangle = \int y^2 P(x, y, t) dx dy. \tag{1.9b}$$

These definitions and Eqs. (1.7) and (1.8) yield subdiffusion in the x direction,

$$\langle x^2(t) \rangle = \mathcal{L}^{-1}\left[\int_{-\infty}^{\infty} g(y) dy \left(-\frac{d^2 f}{dk^2}(k=0,s)\right)\right] = \frac{D_x}{\sqrt{D_y}}\frac{\sqrt{t}}{\Gamma(3/2)}, \tag{1.10}$$

and normal diffusion in the fingers,

$$\langle y^2(t) \rangle = \mathcal{L}^{-1}\left[\int_{-\infty}^{\infty} y^2 g(y) dy f(k=0,s)\right] = 2D_y t. \tag{1.11}$$

Here $\Gamma(z+1) = z\Gamma(z)$ is the gamma function, whose properties are discussed in greater detail in Chapter 2. We also used in Eqs. (1.10) and (1.11) that

$$\tilde{P}(k=0, y, t) = \int_{-\infty}^{\infty} P(x, y, t) dx, \tag{1.12a}$$

$$-\frac{d^2}{dk^2}\tilde{P}(k, y, t)\Big|_{k=0} = \int_{-\infty}^{\infty} x^2 P(x, y, t) dx, \tag{1.12b}$$

which follows from the definition of the Fourier transformation (1.4).

1.2.1 *Fractional Fokker–Planck equation*

Equation (1.11) implies that the comb diffusion equation (1.1) reduces to the normal diffusion equation for $y \neq 0$. The same result is obtained by integrating Eq. (1.1) with respect to x and taking into account the boundary conditions. However, the integration with respect to y leads to a completely different equation for the marginal distribution,

$$P_1(x,t) = \int_{-\infty}^{\infty} P(x,y,t)dy. \tag{1.13}$$

The equation reads

$$\frac{\partial}{\partial t} P_1(x,t) = D_x \frac{\partial^2}{\partial x^2} P(x,y=0,t), \tag{1.14}$$

where $P(x,y=0,t)$ is given by its Laplace transform $\hat{P}(x,y=0,s) = f(x,s)$. To obtain this equation in closed form, we return to Eq. (1.7) in (x,y)-space,

$$\hat{P}(x,y,s) = \hat{P}(x,y=0,s) \exp\left[-\sqrt{s/D_y}|y|\right], \tag{1.15}$$

and integrate it with respect to y. This yields the relation $\hat{P}(x,y=0,s) = \sqrt{s/4D_y}\hat{P}_1(x,s)$. Using this relation and Laplace transforming Eq. (1.14), we find

$$s^{\frac{1}{2}}\hat{P}_1(x,s) - \delta(x)s^{-\frac{1}{2}} = \frac{D_x}{2\sqrt{D_y}} \frac{\partial^2}{\partial x^2} \hat{P}_1(x,s). \tag{1.16}$$

Performing the inverse Laplace transformation, we obtain the integro-differential equation

$$ {}_0^C\mathcal{D}_t^{\frac{1}{2}} P_1(x,t) = \frac{D_x}{2\sqrt{D_y}} \frac{\partial^2}{\partial x^2} P_1(x,t). \tag{1.17}$$

The integro-differential operator ${}_0^C\mathcal{D}_t^{\alpha} \equiv {}^C\mathcal{D}_t^{\alpha}$ is the so-called Caputo fractional derivative,

$$ {}^C\mathcal{D}_t^{\alpha} P_1(t) = \frac{1}{\Gamma(1-\alpha)} \int_0^t \frac{1}{(t-t')^{\alpha}} \frac{dP_1(t')}{dt'} dt', \quad 0 < \alpha < 1, \tag{1.18}$$

while Eq. (1.17) is called the fractional Fokker–Planck equation (FFPE). Instead of the Caputo derivative, it is possible to employ the Riemann–Liouville fractional derivative,

$$ \mathcal{D}_t^{\alpha} P_1(t) \equiv {}^{RL}_0\mathcal{D}_t^{\alpha} P_1(t) = \frac{1}{\Gamma(1-\alpha)} \frac{d}{dt} \int_0^t \frac{P_1(t')dt'}{(t-t')^{\alpha}}, \quad 0 < \alpha < 1, \tag{1.19}$$

which leads to a different form of the FFPE [3],

$$\frac{\partial}{\partial t} P_1(x,t) = \frac{D_x}{2\sqrt{D_y}} \mathcal{D}_t^{1-\alpha} \frac{\partial^2}{\partial x^2} P_1(x,t), \tag{1.20}$$

where $\alpha = 1/2$. This equation can be obtained from Eq. (1.16) by multiplying it by $s^{1/2}$ and Laplace transforming. For arbitrary $0 < \alpha < 1$, this equation is a general form of the FFPE. In that case, the diffusion coefficient must be generalized as well, $D_x/(2\sqrt{D_y}) \to D_{\alpha}$. Fractional derivatives are discussed in greater detail in Chapter 2.

1.2.2 CTRW and Master equation of the comb

We have seen that the transport in the fingers leads to a memory effect in the backbone transport in the form of a fractional integral. The fingers play the role of traps with delay or waiting times that are distributed according to a power law. From the solution (1.7), we obtain that diffusion in the fingers is described by the pdf

$$g(y,t) = \frac{1}{\sqrt{4\pi D_y t}} \exp(-y^2/4D_y t). \tag{1.21}$$

This expression yields the pdf to find a walker at $y = 0$ at time t, $g(0,t) = 1/\sqrt{4\pi D_y t}$. We use this result to obtain an approximate expression for the survival probability[3]

$$\Psi(t) = \int_t^\infty \psi(t')dt' \propto (\tau/\pi t)^{1/2}, \tag{1.22}$$

which implies that the waiting time pdf, or the pdf of the delay times, is $\psi(t) \sim 1/t^{3/2}$, where τ is a characteristic time related to normal diffusion in fingers.

Consequently, as far as the one-dimensional diffusion process along the backbone is concerned, the walker is trapped between any two successive jumps for a random waiting time. These trapping processes are independent and identically distributed. In other words, the transport along the backbone corresponds to a CTRW. This CTRW takes place along the x-direction only, such that the pdf $P_1(x,t)$ is the "genuine" pdf for this one-dimensional diffusion, while it is the marginal pdf for the two-dimensional comb. In the standard CTRW, jumps are also independent and identically distributed. Their pdf is taken to be symmetric with the jump length variance σ^2. In that case, the jumps are described by the second spatial derivative. Therefore, a master equation for the one-dimensional pdf $P_1(x,t)$ can be written as follows,

$$\frac{\partial P_1(x,t)}{\partial t} = \frac{\bar{\sigma}^2}{2} \int_0^t K(t-t') \frac{\partial^2 P_1(x,t')}{\partial x^2} dt', \tag{1.23}$$

where $\bar{\sigma}^2$ is related to the variance of a single jump. The memory kernel $K(t-t')$ is related to the waiting time pdf $\psi(t)$ via its Laplace transform,

$$\hat{K}(s) = s\hat{\psi}(s)/[1 - \hat{\psi}(s)]. \tag{1.24}$$

The Master equation (1.23) and the expression (1.24) for the memory kernel are explained in the following sections.

---

[3]Here we restrict ourself to this rough estimate. An accurate derivation of this expression will be presented in Chapter 3.

1.3   Continuous time random walk

As already mentioned, one of the main assumptions of the CTRW is that the waiting times and jump lengths[4] are independent identically distributed with the pdfs $\psi(t)$ and $w(x)$, respectively. Therefore, realizations of a CTRW result from a sequence of these independent jumps with pdf $w(x)$ and delays with the waiting time pdf $\psi(t)$.

1.3.1   *Renewal theory*

We begin with a discussion of $\psi(t)$. Asking the question how many steps or arrival events N take place during time t with the probability $\mathcal{P}_N(t)$, where the elapsed time t is the sum of the waiting times of the steps, $t = t_1 + t_2 + \cdots + t_N$, we establish a relation between $\mathcal{P}_N(t)$ and $\psi(t)$ using renewal theory [38]. Recognizing that the probability that no events have occurred in the time interval $(0,t)$ is given by

$$\mathcal{P}_{N=0}(t) = 1 - \int_0^t \psi(t')dt' = \int_t^\infty \psi(t')dt' = \Psi(t), \qquad (1.25)$$

we define the function $\Psi(t)$ as the *survival probability*. Clearly,

$$\psi(t) = -\frac{\partial \Psi(t)}{\partial t}. \qquad (1.26)$$

For a single event, $\mathcal{P}_{N=1}(t)$ accounts for the probability that the first event occurs at time t', and there are no subsequent events until time t, which results in the convolution,

$$\mathcal{P}_1(t) = \int_0^t \psi(t')\Psi(t - t')dt' \equiv \psi * \Psi. \qquad (1.27)$$

Since the waiting times are independent, the distribution for two events is obtained by iteration,

$$\mathcal{P}_2(t) = \psi * \mathcal{P}_1(t). \qquad (1.28)$$

This eventually results in a chain of iterations

$$\mathcal{P}_N(t) = \psi * \mathcal{P}_{N-1} = \overbrace{\psi * \psi \cdots * \psi}^{N} *\Psi, \qquad (1.29)$$

where the over-brace contains N convolutions. Equation (1.29) yields also a chain for the pdfs of N steps during time t,

$$\psi_N(t) = \int_0^t \psi_{N-1}(t')\psi(t - t')dt' \equiv \psi * \psi_{N-1}(t) = \overbrace{\psi * \psi \cdots * \psi}^{N} \qquad (1.30)$$

---

[4]An instructive discussion of random walk theory and CTRWs can be found, e.g., in the book by Klafter and Sokolov [37].

The Laplace transforms of $\psi_N(t)$ and $\mathcal{P}_N(t)$ follow from the Laplace transform of integrals. From Eq. (1.25), we have

$$\hat{\Psi}(s) = \int_0^\infty e^{-st}\Psi(t)dt = \frac{1 - \hat{\psi}(s)}{s}, \tag{1.31}$$

where

$$\hat{\psi}(s) = \int_0^\infty e^{-st}\psi(t)dt \equiv \mathcal{L}[\psi(t)](s), \tag{1.32}$$

which also yields

$$\mathcal{L}\left[\overbrace{\psi * \psi \cdots * \psi}^{N}\right](s) = [\hat{\psi}(s)]^N. \tag{1.33}$$

Eventually, we obtain

$$\hat{\mathcal{P}}_N(s) = \mathcal{L}\left[\mathcal{P}_N(t)\right] = [\hat{\psi}(s)]^N \frac{1 - \hat{\psi}(s)}{s}. \tag{1.34}$$

1.3.2 Random steps

Now let us consider only the jumps, which are instantaneous events. In this case, x is the position of a walker after N jumps, $x = \sum_{n=1}^{N} x_n$. It is worth stressing that in this jumps-only process, the number of jumps $n \in [1, N]$ plays the role of an internal clock [37]. We introduce the pdf of being at the position x after N jumps as $w_N(x)$, where $w_1(x) = w(x)$. In complete analogy with the waiting times, it corresponds to the same Markov chain as in Eq. (1.30),

$$w_N(x) = \int_{-\infty}^\infty w_{N-1}(x')w(x - x')dx' \equiv w * w_{N-1} = \overbrace{w * w \cdots * w}^{N}. \tag{1.35}$$

This expression assumes that $x \equiv x_N$ and $x' \equiv x_{N-1}$ are positions of a walker after N and $N - 1$ jumps respectively, where $x, x' \in \mathbb{R}$, and that all possible jump lengths $x - x'$ along the infinite axis are distributed according to $w(x - x')$.

By means of the Fourier transformation we introduce a characteristic function for the N steps,

$$\tilde{w}_N(k) = \mathcal{F}[w_N(x)](k) = \int_{-\infty}^\infty w_N(x)e^{ikx}dx. \tag{1.36}$$

In particular for $N = 1$, it is the average of the exponential,

$$\tilde{w}(k) = \mathcal{F}[w(x)](k) = \int_{-\infty}^\infty e^{ikx}w(x) = \langle e^{ikx} \rangle. \tag{1.37}$$

Applying the Fourier transformation to the chain (1.35) we obtain from the definition (1.36)

$$\tilde{w}_N(k) = \int_{-\infty}^{\infty} e^{ikx} \int_{-\infty}^{\infty} w_{N-1}(x') \frac{1}{2\pi} \int_{-\infty}^{\infty} \tilde{w}(k') e^{-ik'(x-x')} dk' dx' dx$$

$$\int_{-\infty}^{\infty} \tilde{w}_{N-1}(k') \tilde{w}(k') \delta(k'-k) dk' = \tilde{w}_{N-1}(k) \tilde{w}(k) = [\tilde{w}(k)]^N . \quad (1.38)$$

1.3.3   *Montroll–Weiss equation*

We now combine the two considerations of the previous sections and consider a random walk in space and time. To this end we introduce the pdf $\mathcal{P}(x,t)$ to find a particle or a walker at time t at position x. Taking into account all possible N steps with $N \in [0, \infty]$ that a walker made during time t to arrive at the position x, we obtain the pdf $\mathcal{P}(x,t)$ as a subordination

$$\mathcal{P}(x,t) = \sum_{N=0}^{\infty} \mathcal{P}_N(t) w_N(x), \quad (1.39)$$

which translates[5] the internal clock N of the random walk (1.35) to the physical time t. Performing the Fourier and Laplace transformations, we obtain Eq. (1.39) as follows,

$$\hat{\tilde{\mathcal{P}}}(k,s) = \sum_{N=0}^{\infty} \hat{\mathcal{P}}_N(s) \tilde{w}_N(k). \quad (1.40)$$

We replace $\hat{\mathcal{P}}_N(s)$ and $\tilde{w}_N(k)$ by their expressions in Eqs. (1.34) and (1.38),

$$\hat{\tilde{\mathcal{P}}}(k,s) = \frac{1 - \hat{\psi}(s)}{s} \sum_{N=0}^{\infty} [\hat{\psi}(s)]^N [\tilde{w}(k)]^N$$

$$= \frac{1 - \hat{\psi}(s)}{s} \cdot \frac{1}{1 - \tilde{w}(k)\hat{\psi}(s)}. \quad (1.41)$$

This is the well-known Montroll–Weiss equation [5], which is the central result of the CTRW theory. Here it is presented in a simplified decoupled form for a one-dimensional random walk.

1.3.3.1   *Diffusion limit*

We now show that Eq. (1.41) is related to the FFPE (1.17). To this end, we derive Eq. (1.16) from the Montroll–Weiss equation (1.41). First, the

---

[5]A pedagogical explanation of subordination can be found in Ref. [37].

marginal pdf for the FFPE and the CTRW pdf are the same, $P_1(x,t) = P(x,t)$, with the same initial conditions in the form of the $\delta(x)$ function. Second, the FFPE corresponds to an asymptotic equation, and we can obtain it in the diffusion limit $(s,k) \to (0,0)$ from the Montroll–Weiss equation. To do so, we derive the asymptotic form of the Fourier and the Laplace transforms $\tilde{w}(k)$ and $\hat{\psi}(s)$.

We assume that the jump length pdf is a Gaussian,

$$w(x) = (4\pi\sigma^2)^{-1/2}\exp(-x^2/4\sigma^2).$$ (1.42)

The Fourier transform is given by

$$\tilde{w}(k) = \frac{1}{\sqrt{4\pi\sigma^2}}\int_{-\infty}^{\infty} e^{-\frac{x^2}{4\sigma^2}}e^{ikx}\,dx = e^{-\sigma^2 k^2} \approx 1 - \sigma^2 k^2,$$ (1.43)

since $\sigma^2 k^2 \ll 1$ in the diffusion limit. The Laplace transform of the waiting time pdf can be obtained from the asymptotic form of the survival probability (1.22), which is $\Psi(t) \sim (\tau/\pi t)^{1/2}$. Laplace transformation of this expression yields Eq. (1.31) for the Laplace transform of the survival probability, while the right side of the expression yields the gamma function,

$$\Gamma(1/2) = \int_0^{\infty} z^{-\frac{1}{2}}e^{-z}\,dt = \sqrt{\pi}.$$ (1.44)

Therefore we have $[1 - \hat{\psi}(s)]/s \sim (\tau/s)^{1/2}$, which yields

$$\hat{\psi}(s) \approx 1 - (\tau s)^{1/2},$$ (1.45)

since $\tau s \ll 1$ in the diffusion limit. Substituting the asymptotic expressions (1.43) and (1.45) into Eq. (1.41), defining $D_{\frac{1}{2}} = \sigma^2/\tau^{\frac{1}{2}}$, and $\hat{\tilde{P}}(k,s) \to \hat{\tilde{P}}_1(k,s)$ we obtain

$$s^{\frac{1}{2}}\hat{\tilde{P}}_1(k,s) - s^{-\frac{1}{2}} = -D_{\frac{1}{2}}k^2\hat{\tilde{P}}_1(k,s).$$ (1.46)

After the inverse Fourier transformation, we obtain Eq. (1.16). Upon performing the inverse Laplace transformation, we obtain the integro-differential equation (1.17).

1.3.3.2 *Master equation*

Next we explain the memory kernel $K(t)$ in the Master equation (1.23) and its relation to the waiting time pdf in Laplace space,

$$\hat{K}(s) = s\hat{\psi}(s)/[1 - \hat{\psi}(s)].$$ (1.47)

It is customary to derive the Master equation from a balance equation, or the Liouville continuity equation, see for example [37]. However, we already have the Montroll–Weiss equation as the main equation for CTRWs. Therefore, we show that the desired result also follows from Eq. (1.41). Let us rewrite it as follows,

$$s\hat{\tilde{P}}(k,s)\left[1+\frac{\hat{\psi}(s)}{1-\hat{\psi}(s)}\right]-1=\frac{s\tilde{w}(k)\hat{\psi}(s)}{1-\hat{\psi}(s)}\hat{\tilde{P}}(k,s). \qquad (1.48)$$

Here we used that

$$1/[1-\hat{\psi}(s)]=1+\hat{\psi}(s)/[1-\hat{\psi}(s)]. \qquad (1.49)$$

From this we obtain the desired form of the Master equation in Fourier–Laplace space,

$$s\hat{\tilde{P}}(k,s)-1=\hat{K}(s)[\tilde{w}(x)-1]\hat{\tilde{P}}(k,s), \qquad (1.50)$$

where the memory kernel is defined in Eq. (1.47). Performing the inverse Fourier and Laplace transformations, we obtain the Master equation in a general form, including for the three dimensional random walk $x \to \mathbf{r} \in \mathbb{R}^3$. However for our one-dimensional Gaussian jumps, when $\tilde{w}(k) - 1 \approx -\sigma^2 k^2$, we obtain the Master equation (1.23) with $\bar{\sigma}^2 = 2\sigma^2$ [3].

Chapter 2

Fractional calculus and special functions

"Fractional calculus is the field of mathematical analysis which deals with the investigation and applications of integrals and derivatives of arbitrary order. The term fractional is a misnomer, but it is retained following the prevailing use. The fractional calculus may be considered an *old* and yet *novel topic*. It is an old topic since, starting from some speculations of G.W. Leibniz (1695, 1697) and L. Euler (1730), it has been developed up to nowadays. ...

However, it may be considered a novel topic as well, since only from a little more than [twenty] years it has been object of specialized conferences and treatises. ..."

Prof. R. Gorenflo and Prof. F. Mainardi
in "Fractional Calculus" (1991)

Fractional calculus was inspired in 1695 by L'Hopital's query in a letter to Leibniz about a fractional exponent ν in derivative $d^\nu y/dx^\nu$ of a function $y(x)$. During the next 200 years, this topics was mentioned from time to time by eminent mathematicians such as Euler, Lagrange, and Fourier. The most significant contributions to the topic were made by Abel and Liouville. The modern theory of fractional calculus was started by Riemann. The textbook by Oldham and Spanier [39] presents a detailed, comprehensive chronology of the mathematical results. A historical survey with interesting mathematical examples can also be found in a monograph by Miller and Ross [40].

The objective of this chapter is not to provide a comprehensive description of the various aspects of fractional calculus. Those are covered thoroughly with mathematical rigor by excellent monographs, such as Refs. [39–42]. Our main focus here is to provide a concise description of some essential tools and techniques of fractional calculus. We also present some required details on special functions that are commonly used in fractional calculus and that often appear in fractional kinetics. This basic information is needed to understand the treatment of fractional differential equations, which describe various realizations of anomalous transport in combs studied in this monograph.

We start with the gamma and beta functions, $\Gamma(\nu)$ and $B(\mu, \nu)$, whose properties play a crucial role in fractional integration.[1]

2.1   Gamma and Beta functions

The gamma function is defined for complex numbers with a positive real part by the integral

$$\Gamma(\alpha) = \int_0^\infty x^{\alpha-1} e^{-x} dx, \quad \alpha > 0. \tag{2.1}$$

It is extended by analytical continuation to all complex numbers, except the non-positive integers, where it has simple poles, see below. Integrating by part, we obtain a recursion formula,

$$\Gamma(\alpha + 1) = \alpha \Gamma(\alpha). \tag{2.2}$$

An alternative representation of the gamma function is via Euler's limit,

$$\Gamma(\alpha) = \lim_{j \to \infty} \left[\frac{j! j^\alpha}{\alpha(\alpha+1) \dots (\alpha+j)} \right], \tag{2.3}$$

where $\Gamma(1) = 1$. For the integer $\alpha = n$, the recursion (2.2) yields $\Gamma(n+1) = n\Gamma(n) = n!$. Recursion (2.2) also defines the gamma function for negative arguments as an analytical continuation $\Gamma(z-1) = \Gamma(z)/(z-1)$, where $\Gamma(0)$ and $\Gamma(-n)$ diverge, while the ratio is finite $\Gamma(-n)/\Gamma(-m) = m!/n!$.

Example 2.1. Let us calculate $\Gamma(1/2) = (-1/2)!$:

$$\Gamma(1/2) = \int_0^\infty x^{-1/2} e^{-x} dx = 2 \int_0^\infty e^{-y^2} dy = \sqrt{\pi}. \tag{2.4}$$

$\square$

Another useful function is the beta function with the integral representation for $p, q > 0$,

$$B(p, q) = \int_0^1 x^{p-1} (1-x)^{q-1} dx, \tag{2.5}$$

---

[1] We follow Refs. [43] and [41] here.

while for arbitrary p and q, it is defined as a composition of gamma functions,

$$B(p,q) = \frac{\Gamma(p)\Gamma(q)}{\Gamma(p+q)}. \tag{2.6}$$

Using the beta function, some additional properties of the gamma function can be established as well, as shown in the following examples.

Example 2.2. To prove formula (2.6) [43], we substitute $x = y/(1+y)$ in the definition (2.5) and obtain another integral representation of the beta function,

$$B(p,q) = \int_0^\infty y^{p-1}(1+y)^{-p-q}dy. \tag{2.7}$$

Next we establish the following identity,

$$\int_0^\infty x^{p+q-1}e^{-(1+y)x}dx = \frac{\Gamma(p+q)}{(1+y)^{p+q}}. \tag{2.8}$$

Carrying out the variable substitution $z = (1+y)x$, we find

$$\int_0^\infty x^{p+q-1}e^{-(1+y)x}dx = \int_0^\infty \left(\frac{z}{1+y}\right)^{p+q-1} e^{-z}\frac{dz}{1+y}$$

$$= \frac{1}{(1+y)^{p+q}}\int_0^\infty z^{p+q-1}e^{-z}dz, \tag{2.9}$$

which leads to the relation (2.8), using the definition (2.1). We multiply Eq. (2.8) by y^{p-1} and integrate with respect to y. Changing the order of integration, we find

$$\int_0^\infty dx\, x^{p+q-1}e^{-x}\int_0^\infty y^{p-1}e^{-xy}dy$$

$$= \int_0^\infty dx\, x^{p+q-1}e^{-x}\frac{\Gamma(p)}{x^p}$$

$$= \Gamma(p)\int_0^\infty dx\, x^{q-1}e^{-x} = \Gamma(p+q)\int_0^\infty y^{p-1}(1+y)^{-p-q}dy, \tag{2.10}$$

where we have used the identity (2.8) to evaluate the second integral. Taking into account definitions (2.1) and (2.7), we obtain Eq. (2.6). $\qquad\square$

Example 2.3. Consider both expressions (2.5) and (2.6) for $q = 1 - p$:

$$\Gamma(p)\Gamma(1-p) = B(p, 1-p) = \int_0^1 \left(\frac{x}{1-x}\right)^{p-1}\frac{dx}{1-x}. \tag{2.11}$$

The integral converges for $0 < p < 1$. Making the change of the variable $y = x/(1 - x)$, which yields $dx/(1 - x) = dy/(1 + y)$, we obtain

$$\Gamma(p)\Gamma(1 - p) = \int_0^\infty \frac{y^{p-1}dy}{1 + y}. \tag{2.12}$$

We perform the analytical continuation in the complex plane with $y \to z = re^{i\phi}$ and with a branch cut along the real axis $(0, \infty)$, resulting in the contour integral

$$I = \oint_C \frac{z^{p-1}dz}{1 + z}. \tag{2.13}$$

The contour of the integration C consists of four parts: (1) the integration over a circle with large radius $r = R \to \infty$, (2) the integration over a circle around $z = 0$ with small radius $r = \epsilon \to 0$, and (3) two integrals, one over the upper edge of the cut along the real axis (ϵ, R) and the second over lower edge of the cut in the interval (R, ϵ). There is only a simple pole at $z = e^{i\pi}$, which yields $I = 2\pi i e^{i\pi p}$, according to the residue theorem. We have

$$I = I_R + I_\epsilon + I_{(\epsilon,R)}(\phi = 0) + I_{(R,\epsilon)}(\phi = 2\pi) = 2\pi i e^{\pi i p}, \tag{2.14}$$

and for the individual integrals,

$$I_{R=\infty} = I_{\epsilon=0} = 0, \tag{2.15}$$
$$I_{0,\infty} = B(p, 1 - p), \tag{2.16}$$
$$I_{\infty,0} = -e^{2\pi i p}B(p, 1 - p). \tag{2.17}$$

Taking all these results into account, we finally obtain Euler's reflection formula,

$$\Gamma(p)\Gamma(1 - p) = \frac{\pi}{\sin(p\pi)}. \tag{2.18}$$

$\square$

Equation (2.18) implies that $\Gamma(1/2) = \sqrt{\pi}$, see also Example 2.1.

Example 2.4. Another property that follows immediately from the beta function (2.5) is the Legendre duplication formula. From the integral (2.5), we have

$$B(p, p) = \int_0^1 [\tau(1 - \tau)]^{p-1}d\tau. \tag{2.19}$$

Performing the change of the variable $s = 4\tau(1-\tau)$, which is symmetric with respect to $\tau = 1/2$, we find

$$B(p,p) = 2 \int_0^{\frac{1}{2}} [\tau(1-\tau)]^{p-1} d\tau$$

$$= \frac{1}{2^{2p-1}} \int_0^1 s^{p-1}(1-s)^{-1/2} ds = \frac{1}{2^{2p-1}} B(p, 1/2). \quad (2.20)$$

Using the definition (2.6) and the fact that $\Gamma(1/2) = \sqrt{\pi}$, we obtain the Legendre duplication formula,

$$\Gamma(p)\Gamma(p+1/2) = 2^{1-2p}\sqrt{\pi}\,\Gamma(2p), \quad 2p \neq -1. \quad (2.21)$$

$\square$

2.1.1 *Contour integral representation of* $\Gamma(p)$ *and* $1/\Gamma(p)$

We perform the analytical continuation in the complex plane $z = x - iy$ in the integration (2.1) to represent it as a contour integral

$$\int_C z^{p-1} e^{-z} dz = \int_C e^{(p-1)[\ln(z)]-z} dz, \quad (2.22)$$

where the contour C starts at $+\infty$, runs around the point $z = 0$, and ends again at $+\infty$. Since $z = 0$ is a branch point, we take a branch cut along the non-negative real axis $(0, \infty)$. In this case, the contour consists of three parts: (1) the upper edge of the branch cut (∞, ϵ), (2) the circle C_ϵ of radius $\epsilon \to 0$ with the center at $z = 0$, and (3) the lower edge of the branch cut (ϵ, ∞). On the upper edge, $\ln z = \ln x$ is real, while on the lower edge $\ln z = \ln x + 2\pi i$. Therefore,

$$\int_C z^{p-1} e^{-z} dz = \int_\infty^\epsilon x^{p-1} e^{-x} dx + \int_{C_\epsilon} z^{p-1} e^{-z} dz + e^{2(p-1)\pi i} \int_\epsilon^\infty x^{p-1} e^{-x} dx.$$

$$(2.23)$$

In the limit $\epsilon \to 0$, integration over the contour C_ϵ vanishes, and we obtain

$$\Gamma(p) = \int_0^\infty x^{p-1} e^{-x} dx = \frac{1}{e^{2(p-1)\pi i} - 1} \int_C z^{p-1} e^{-z} dz. \quad (2.24)$$

To obtain a formula for the contour integral representation of $1/\Gamma(p)$, we replace p by $1-p$ in Eq. (2.24),

$$\int_C z^{-p} e^{-z} dz = (e^{-2p\pi i} - 1)\Gamma(1-p). \quad (2.25)$$

With the further substitution $z = \tau e^{\pi i} = -\tau$, we invert the contour C with respect to the y axis. In this case, the contour is known as a Hankel contour $\{Ha\}$. The contour integral (2.25) reads

$$\int_C z^{-p} e^{-z} dz = -e^{-p\pi i} \int_{\{Ha\}} \tau^{-p} e^\tau d\tau. \quad (2.26)$$

Combining Eqs. (2.25) and (2.18), we find

$$\int_{\{Ha\}} \tau^{-p} e^{\tau} d\tau = 2i \sin(p\pi)\Gamma(1-p) = \frac{2\pi i}{\Gamma(p)}. \qquad (2.27)$$

This implies the following integral representation of the reciprocal gamma function,[2]

$$\frac{1}{\Gamma(p)} = \frac{1}{2\pi i} \int_{\{Ha\}} \tau^{-p} e^{\tau} d\tau. \qquad (2.28)$$

2.2   Mittag–Leffler function

Mittag–Leffler functions play a crucial role in solving fractional kinetic equation. They are generalizations of the exponential function e^z and can be defined by means of the gamma function $\Gamma(p)$ with $p = k\alpha + \beta$, where $\alpha, \beta > 0$ and k is an integer. A two-parameter Mittag–Leffler function is defined by the series

$$E_{\alpha,\beta}(z) = \sum_{k=0}^{\infty} \frac{z^k}{\Gamma(k\alpha + \beta)}, \quad \alpha > 0, \ \beta > 0. \qquad (2.29)$$

Example 2.5. As follows from the definition (2.29),

$$E_{\alpha,1}(z) \equiv E_{\alpha}(z) = \sum_{k=0}^{\infty} \frac{z^k}{\Gamma(k\alpha + 1)} \qquad (2.30)$$

is a one parameter generalization of the exponential function, where

$$E_1(z) = \sum_{k=0}^{\infty} \frac{z^k}{\Gamma(k+1)} = e^z, \qquad (2.31)$$

$$E_2(z^2) = \sum_{k=0}^{\infty} \frac{z^{2k}}{\Gamma(2k+1)} = \sum_{k=0}^{\infty} \frac{z^{2k}}{2k!} = \cosh(z), \qquad (2.32)$$

$$E_{\frac{1}{2}}(z) = \sum_{k=0}^{\infty} \frac{z^k}{\Gamma(\frac{k}{2}+1)} = e^{z^2} \mathrm{erfc}(-z). \qquad (2.33)$$

$\square$

---

[2]By deformation of the contour, this expression is the inverse Laplace transform.

We substitute the integral representation of the reciprocal gamma function (2.27) into the definition (2.29),

$$E_{\alpha,\beta}(z) = \sum_{k=0}^{\infty} \frac{z^k}{2\pi i} \int_{\{Ha\}} \tau^{-k\alpha-\beta} e^\tau d\tau = \frac{1}{2\pi i} \int_C \frac{\tau^{\alpha-\beta} e^\tau d\tau}{\tau^\alpha - z}. \qquad (2.34)$$

We have obtained an integral representation of the Mittag–Leffler function, where the second equality follows from an expansion of $1/(\tau^\alpha - z)$ in powers of z and integrating term-by-term [43]. Here C is Hankel's contour deformed to a loop that starts and ends at $\tau = -\infty$, encircling a disk $|\tau| < |z|^{1/\alpha}$ in the positive direction [43]. In practical terms, we have arrived at the definition of the Mittag–Leffler function in the form of the inverse Laplace transform. To see this, we introduce new variables $\tau = st$ and $z = \lambda t^\alpha$, which yields

$$E_{\alpha,\beta}(\lambda t^\alpha) = \frac{t^{1-\beta}}{2\pi i} \int_C \frac{s^{\alpha-\beta} e^{st} ds}{s^\alpha - \lambda} \equiv t^{1-\beta} \mathcal{L}^{-1}\left[\frac{s^{\alpha-\beta}}{s^\alpha - \lambda}\right]. \qquad (2.35)$$

Example 2.6. Equation (2.35) is convenient for the integration of the Mittag–Leffler function. In particular, we have [41]

$$\int_0^\infty t^{\beta-1} e^{-t} E_{\alpha,\beta}(\lambda t^\alpha)\, dt = \frac{1}{1-\lambda}. \qquad (2.36)$$

$\square$

2.2.1 *Asymptotic properties*

An important property of the Mittag–Leffler function (2.34) is its asymptotic behavior. Substituting the following expansion

$$\frac{s^\alpha}{s^\alpha + \lambda t^\alpha} = \sum_{n=1}^{N-1} \frac{s^{n\alpha}}{(\lambda t^\alpha)^n} + \left(1 - \frac{s^\alpha}{\lambda t^\alpha}\right)^{-1} \frac{s^{\alpha N}}{\lambda^N t^{\alpha N}}, \qquad (2.37)$$

while neglecting the terms $O\left(t^{-\alpha N}\right)$ for $t \to \infty$, into the contour integral in Eq. (2.34), we obtain

$$E_{\alpha,\beta}(-\lambda t^\alpha) = \sum_{n=1}^{N-1} (\lambda t^\alpha)^{-n} \frac{1}{2\pi i} \int_C \tau^{n\alpha-\beta} e^\tau d\tau$$

$$= \sum_{n=1}^{N-1} \frac{(\lambda t^\alpha)^{-n}}{\Gamma(\beta - n\alpha)} \sim \frac{t^{-\alpha}}{\lambda\Gamma(\beta - \alpha)} \quad \text{for } t\lambda^{1/\alpha} \gg 1, \qquad (2.38)$$

where we used Eq. (2.28).

2.3 Fox H-functions

Another ubiquitous function in the theory of fractional differential equations is the Fox H function (Fox function). A detailed description of the Fox function and its application can be found in Refs. [44, 45]. The Fox H function is defined in terms of the Mellin–Barnes integral,

$$
H_{p,q}^{m,n}(z) = H_{p,q}^{m,n}\left[z \left|\begin{array}{l}(a_p, A_p)\\(b_q, B_q)\end{array}\right.\right]
$$

$$
= H_{p,q}^{m,n}\left[z \left|\begin{array}{l}(a_1, A_1), \ldots, (a_p, A_p)\\(b_1, B_1), \ldots, (b_q, B_q)\end{array}\right.\right] = \frac{1}{2\pi i}\int_C ds\, \theta(s) z^{-s}.
$$

(2.39)

Here

$$
\theta(s) = \frac{\prod_{j=1}^{m}\Gamma(b_j + B_j s)\prod_{j=1}^{n}\Gamma(1 - a_j - A_j s)}{\prod_{j=m+1}^{q}\Gamma(1 - b_j - B_j s)\prod_{j=n+1}^{p}\Gamma(a_j + A_j s)},
$$

(2.40)

where n, m, p, q are non-negative integers with $0 \leq n \leq p$, $1 \leq m \leq q$, $a_i, b_j \in \mathbb{C}$, $A_i, B_j \in \mathbb{R}^+$, $i = 1, \ldots, p$, and $j = 1, \ldots, q$. If the product is empty, then it is set equal to one. The contour C starts at $c - i\infty$, ends at $c + i\infty$, and separates the poles $\xi_{j,k} = -(b_j + k)/B_j$ of the gamma function $\Gamma(b_j + B_j s)$ with $j = 1, \ldots, m$ and $k = 0, 1, 2, \ldots$ from the poles $\chi_{i,k} = (1 - a_i + k)/A_i$ of the function $\Gamma(1 - a_i - A_i s)$, $i = 1, \ldots, n$.

Let us consider some properties of the Fox H-function, which follow immediately from the definition in Eqs. (2.39) and (2.40). The first one is a scaling of the argument,

$$
H_{p,q}^{m,n}\left[z^{\delta} \left|\begin{array}{l}(a_p, A_p)\\(b_q, B_q)\end{array}\right.\right] = \frac{1}{\delta} H_{p,q}^{m,n}\left[z \left|\begin{array}{l}(a_p, A_p/\delta)\\(b_q, B_q/\delta)\end{array}\right.\right],
$$

(2.41)

where $\delta > 0$. The second is the reduction formula,

$$
H_{p,q}^{m,n}\left[z \left|\begin{array}{l}(a_1, A_1), \ldots, (a_{p-1}, A_{p-1}), (b_1, B_1)\\(b_1, B_1), (b_2, B_2), \ldots, (b_q, B_q)\end{array}\right.\right]
$$

$$
= H_{p-1,q-1}^{m-1,n}\left[z \left|\begin{array}{l}(a_1, A_1), \ldots, (a_{p-1}, A_{p-1})\\(b_2, B_2), \ldots, (b_q, B_q)\end{array}\right.\right].
$$

(2.42)

where $m \geq 1$, and $p > n$. In the next examples we show how the Fox H-function relates to the exponential function and the Mittag–Leffler function [44].

Example 2.7. We evaluate the integration of the gamma function via a sum of residues at the poles $s = 0, -1, -2, \ldots$,

$$H_{0,1}^{1,0}\left[z\,\middle|\,{(0,1)}\right] = \frac{1}{2\pi i}\int_{c-i\infty}^{c+i\infty}\Gamma(s)z^{-s}ds = \sum_{k=0}^{\infty}\lim_{s\to-k}(s+k)\Gamma(s)z^{-s}$$

$$= \sum_{k=0}^{\infty}\lim_{s\to-k}\frac{(s+k)(s+k-1)\cdots s}{(s+k-1)\cdots s}\Gamma(s)z^{-s}$$

$$= \sum_{k=0}^{\infty}\lim_{s\to-k}\frac{\Gamma(s+k+1)}{(s+k-1)\cdots s}z^{-s} = \sum_{k=0}^{\infty}\frac{(-1)^k}{k!}z^k = e^{-z}.$$

$$(2.43)$$

Again, evaluating the Mellin–Barnes integral as a sum of residues, we have

$$H_{2,1}^{1,1}\left[-z\,\middle|\,{(0,1) \atop (0,1),(1-\beta,\alpha)}\right] = \frac{1}{2\pi i}\int_{c-i\infty}^{c+i\infty}\frac{\Gamma(s)\Gamma(1-s)}{\Gamma(\beta-\alpha s)}(-z)^{-s}ds$$

$$= \sum_{k=0}^{\infty}\lim_{s\to-k}\frac{(s+k)\pi}{\sin(\pi s)\Gamma(\beta-\alpha s)}(-z)^{-s} = \sum_{k=0}^{\infty}\frac{z^k}{\Gamma(k\alpha+\beta)} = E_{\alpha,\beta}(z). \quad (2.44)$$

Here we have used the limit $\lim_{s\to-k}[(s+k)\pi]/\sin(s\pi) = \lim_{s\to-k}(-1)^k[(s+k)\pi]/\sin[(s+k)\pi]$ and the property (2.18).

Using the same technique of a sum of residues, we also establish a relation between the Fox H-function and the modified Bessel function. Let us consider the Mellin–Barnes integral

$$H_{0,2}^{2,0}\left[\frac{z^2}{4}\,\middle|\,{\left(\frac{a+\nu}{2},1\right),\left(\frac{a-\nu}{2},1\right)}\right]$$

$$= \frac{1}{2\pi i}\int_{c-i\infty}^{c+i\infty}\Gamma\left(\frac{a+\nu}{2}+s\right)\Gamma\left(\frac{a-\nu}{2}+s\right)\left(\frac{z}{2}\right)^{-2s}ds$$

$$= \left(\frac{z}{2}\right)^a\left[\sum_{k=0}^{\infty}\lim_{s+\frac{\nu}{2}\to-k}\left(s+\frac{\nu}{2}+k\right)\Gamma\left(\frac{\nu}{2}+s\right)\Gamma\left(\frac{-\nu}{2}+s\right)\left(\frac{z}{2}\right)^{-2s}\right.$$

$$+ \sum_{k=0}^{\infty}\lim_{s-\frac{\nu}{2}\to-k}\left(s-\frac{\nu}{2}+k\right)\Gamma\left(\frac{\nu}{2}+s\right)\Gamma\left(\frac{-\nu}{2}+s\right)\left(\frac{z}{2}\right)^{-2s}\right]$$

$$= \left(\frac{z}{2}\right)^a\left[\sum_{k=0}^{\infty}\lim_{s+\frac{\nu}{2}\to-k}\frac{\Gamma\left(s+k+\frac{\nu}{2}+1\right)}{(s+k-1+\nu/2)\cdots(s+\nu/2)}\Gamma\left(\frac{-\nu}{2}+s\right)\left(\frac{z}{2}\right)^{-2s}\right.$$

$$+ \sum_{k=0}^{\infty}\lim_{s-\frac{\nu}{2}\to-k}\frac{\Gamma\left(s+k-\frac{\nu}{2}+1\right)}{(s+k-1-\nu/2)\cdots(s-\nu/2)}\Gamma\left(\frac{\nu}{2}+s\right)\left(\frac{z}{2}\right)^{-2s}\right]$$

$$= \left(\frac{z}{2}\right)^a\frac{\pi}{\sin(\nu\pi)}\left[\left(\frac{z}{2}\right)^{-\nu}\sum_{k=0}^{\infty}\frac{\left(\frac{z}{2}\right)^{2k}}{k!\Gamma(k-\nu+1)} - \left(\frac{z}{2}\right)^{\nu}\sum_{k=0}^{\infty}\frac{\left(\frac{z}{2}\right)^{2k}}{k!\Gamma(k+\nu+1)}\right]$$

$$= \left(\frac{z}{2}\right)^a \frac{\pi}{\sin(\nu\pi)}[I_{-\nu}(z) - I_\nu(z)] = 2\left(\frac{z}{2}\right)^a K_\nu(z), \tag{2.45}$$

where $I_\nu(z)$ and $K_\nu(z)$ are the modified Bessel functions [46].   □

The Mellin transform is defined as

$$\mathcal{M}[f(x)](s) = \int_0^\infty f(x)x^{s-1}dx, \tag{2.46}$$

and the contour integral in definition (2.39) is the inverse Mellin transform. Therefore, the Mellin transform of the Fox H-function yields

$$\int_0^\infty dx\, x^{s-1} H_{p,q}^{m,n}\left[ax \left|\begin{array}{c} (a_p, A_p) \\ (b_q, B_q) \end{array}\right.\right] = a^{-s}\theta(s), \tag{2.47}$$

where $\theta(s)$ is defined in Eq. (2.40).

Example 2.8. The integral

$$I = \int_0^\infty x^{\alpha-1} e^{-ax-bx^{-\rho}}dx, \tag{2.48}$$

where $a, b, \rho > 0$, can be represented in terms of a Fox H-function, see, e.g., Ref. [44, p. 10]. Here we present an alternative way of evaluating the integral. Noting that the integral (2.48) can be understood as the Laplace transform

$$I = I(a) = \mathcal{L}[x^{\alpha-1}e^{-bx^{-\rho}}](a), \tag{2.49}$$

we perform the following chain of integrations by means of the Mellin transform,

$$I(a) \equiv \mathcal{M}^{-1}\left[\mathcal{M}[I](\zeta)\right](a)$$

$$= \frac{1}{2\pi i}\int_C d\zeta a^{-\zeta}\int_0^\infty da\, a^{\zeta-1}\int_0^\infty x^{\alpha-1}e^{-ax-bx^{-\rho}}dx$$

$$= \frac{1}{2\pi i}\int_C d\zeta a^{-\zeta}\Gamma(\zeta)\int_0^\infty x^{\alpha-\zeta-1}e^{-bx^{-\rho}}dx$$

$$= \frac{\rho^{-1}}{2\pi i}\int_C d\zeta a^{-\zeta}\Gamma(\zeta)b^{\frac{\alpha}{\rho}-\frac{\zeta}{\rho}}\Gamma\left(\frac{\alpha}{\rho}-\frac{\zeta}{\rho}\right)$$

$$= \frac{(a^\alpha\rho)^{-1}}{2\pi i}\int_C ds\left(ab^{\frac{1}{\rho}}\right)^{-s}\Gamma(\alpha+s)\Gamma(s/\rho)$$

$$= \frac{1}{\rho a^\alpha}H_{0,2}^{2,0}\left[ab^{\frac{1}{\rho}} \left|\begin{array}{c} \\ (\alpha,1), \left(0,\frac{1}{\rho}\right) \end{array}\right.\right]. \tag{2.50}$$

   □

The Laplace transformation of the modified rhs of Eq. (2.39) yields

$$\mathcal{L}\left[\frac{x^{\rho-1}}{2\pi i}\int_C ds\,\theta(s)(ax^\sigma)^{-s}\right](\zeta) = \frac{1}{2\pi i}\int_C ds\,\theta(s)a^{-s}\int_0^\infty e^{-\zeta x}x^{\rho-s\sigma-1}dx$$

$$= \frac{1}{2\pi i}\int_C ds\,\theta(s)a^{-s}\zeta^{s\sigma-\rho}\Gamma(\rho-s\sigma).$$

$$(2.51)$$

Taking this expression into account, we obtain the following Laplace transform formula for the Fox H-function:

$$\mathcal{L}\left\{t^{\rho-1}H_{p,q}^{m,n}\left[at^\sigma\,\middle|\,\begin{matrix}(a_p,A_p)\\(b_q,B_q)\end{matrix}\right]\right\}(s)$$

$$= s^{-\rho}H_{p+1,q}^{m,n+1}\left[as^{-\sigma}\,\middle|\,\begin{matrix}(1-\rho,\sigma),(a_p,A_p)\\(b_q,B_q)\end{matrix}\right]. \qquad (2.52)$$

However, from Eq. (2.28) it follows that the inverse Laplace transformation reads

$$\mathcal{L}^{-1}\left\{s^{-\rho}H_{p,q}^{m,n}\left[as^\sigma\,\middle|\,\begin{matrix}(a_p,A_p)\\(b_q,B_q)\end{matrix}\right]\right\} = t^{\rho-1}H_{p+1,q}^{m,n}\left[\frac{a}{t^\sigma}\,\middle|\,\begin{matrix}(a_p,A_p),(\rho,\sigma)\\(b_q,B_q)\end{matrix}\right].$$

$$(2.53)$$

Example 2.9. As another important example, we consider the Mellin-cosine transform of the Fox H-function discussed in Ref. [44],

$$\int_0^\infty dx\,x^{\rho-1}\cos(\kappa x)H_{p,q}^{m,n}\left[ax^\delta\,\middle|\,\begin{matrix}(a_p,A_p)\\(b_q,B_q)\end{matrix}\right]$$

$$= \frac{\sqrt{\pi}2^{\rho-1}}{\kappa^\rho}H_{p+2,q}^{m,n+1}\left[a\left(\frac{2}{\kappa}\right)^\delta\,\middle|\,\begin{matrix}\left(\dfrac{2-\rho}{2},\dfrac{\delta}{2}\right),(a_p,A_p),\left(\dfrac{1-\rho}{2},\dfrac{\delta}{2}\right)\\(b_q,B_q)\end{matrix}\right],$$

$$(2.54)$$

where

$$\Re\left(\rho+\delta\min_{1\le j\le m}\left(\frac{b_j}{B_j}\right)\right) > 0, \quad \kappa^\delta > 0, \qquad (2.55)$$

$$\Re\left(\rho+\delta\max_{1\le j\le n}\left(\frac{a_j-1}{A_j}\right)\right) < 1, \quad |\arg(a)| < \pi\alpha/2, \qquad (2.56)$$

$$\alpha = \sum_{j=1}^n A_j - \sum_{j=n+1}^p A_j + \sum_{j=1}^m B_j - \sum_{j=m+1}^q B_j > 0. \qquad (2.57)$$

The definitions (2.39) and (2.40) of the Fox H-function lead to the Fourier-cosine transform of a power law function,

$$\int_0^\infty x^{\rho-s\delta-1}\cos(\kappa x)dx = \frac{\pi\kappa^{s\delta-\rho}}{2\Gamma(s\delta+1-\rho)}\cos\left(\pi\frac{s\delta+1-\rho}{2}\right). \quad (2.58)$$

Taking into account the properties of the gamma functions (2.18) and (2.21), we find

$$\frac{\pi}{\cos(\pi\nu)} = \Gamma\left(\frac{1+\nu}{2}\right)\Gamma\left(\frac{1-\nu}{2}\right), \quad (2.59)$$

where $\nu = s\delta + 1 - \rho$, and

$$\frac{1}{\Gamma(2\nu)} = \frac{2^{1-2\nu}\sqrt{\pi}}{\Gamma(\nu)\Gamma(\nu+1/2)}. \quad (2.60)$$

Substituting these expressions into the definition of the Fox function, we obtain the result in Eq. (2.54). $\qquad\square$

2.4 Fractional integration

Let us consider the Cauchy integral to define the n-th derivative by means of a contour integral [40],

$$\frac{d^n}{dz^n}f(z) \equiv D^n f(z) = \frac{n!}{2\pi i}\int_C \frac{f(x)}{(x-z)^{n+1}}dx, \quad (2.61)$$

where C is a suitable contour around z and $f(z)$ is a test function. We assume that the test function satisfies all necessary conditions for the integral (2.61), such as being single-valued and analytical, and is suitable for the further description of fractional integro-differentiation. The residue theorem ensures the existence of the n-th derivative of $f(z)$ at point z. Replacing n by an arbitrary number, say ν, the point z is a branch point, and we take a branch cut along the real axis as we did in the case of the gamma function. We replace the factorial $n!$ by $\nu! = \Gamma(\nu+1)$, and the contour integral has a branch point at $z = x$.

We define the integral $_aD_x^\nu f(x)$, where the subscripts define the limit of integration and ν is the order of integral, by

$$_aD_x^\nu f(x) = \frac{\Gamma(\nu+1)}{2\pi i}\int_a^x (\zeta-x)^{-\nu-1}f(\zeta)d\zeta$$

$$= \frac{\Gamma(\nu+1)}{2\pi i}\int_C (\zeta-x)^{-\nu-1}f(\zeta)d\zeta, \quad (2.62)$$

where the last expression is an analytical continuation in the complex plane $\zeta \to z = x + iy$. C is a contour around x that starts and ends at $\zeta = a$. Cutting the real axis from $x - 0$, we obtain

$$(\zeta - x)^{-\nu-1} = e^{-(\nu+1)[\ln|\zeta - x| + i\phi]}, \tag{2.63}$$

where $\phi = \arg(\zeta - x)$. For the upper edge of the cut, $\phi = \pi$, while $\phi = -\pi$ for the lower edge of the cut. The contour integral (2.62) consists of three parts,

$$\int_C (\zeta - x)^{-\nu-1} f(\zeta) d\zeta = \int_{C(\phi=\pi)} + \int_{C(\phi=-\pi)} + \int_{C_\epsilon}$$

$$= e^{i(\nu+1)\pi} \int_a^{x-\epsilon} (x - x')^{-\nu-1} f(x') dx'$$

$$+ e^{-i(\nu+1)\pi} \int_{x-\epsilon}^a (x - x')^{-\nu-1} f(x') dx'$$

$$+ \int_{C_\epsilon} (\zeta - x)^{-\nu-1} f(\zeta) d\zeta. \tag{2.64}$$

The last integral vanishes in the limit $\epsilon \to 0$, and we obtain

$$_a\mathcal{D}_x^\nu f(x) = \frac{\Gamma(\nu+1)}{2\pi i} \int_C (\zeta - x)^{-\nu-1} f(\zeta) d\zeta$$

$$= \frac{\Gamma(\nu+1) \sin[(\nu+1)\pi]}{\pi i} \int_a^{x-\epsilon} (x - x')^{-\nu-1} f(x') dx'. \tag{2.65}$$

Using Euler's reflection formula (2.18) in the form $\Gamma(1+\nu)\Gamma(-\nu) = \pi/\sin[(\nu+1)\pi]$ and performing the reflection $\nu \to -\nu$, we have

$$_a\mathcal{D}_x^{-\nu} f(x) = \frac{1}{\Gamma(\nu)} \int_a^{x-\epsilon} (x - x')^{\nu-1} f(x') dx'. \tag{2.66}$$

This expression is known as a Riemann–Liouville fractional integral with $\nu > 0$. The lower limit a is arbitrary. For $a = -\infty$, this integral is known as a Weyl integral [39].

Example 2.10. The integration $_0\mathcal{D}_t^{-\nu} t^\mu$ with $\mu > -1$ yields:

$$_0\mathcal{D}_t^{-\nu} t^\mu = \frac{1}{\Gamma(\nu)} \int_0^t (t - \tau)^{\nu-1} \tau^\mu d\tau = t^{\mu+\nu} \frac{B(\nu, \mu+1)}{\Gamma(\nu)} = \frac{\Gamma(\mu+1)}{\Gamma(\mu+\nu+1)} t^{\mu+\nu}. \tag{2.67}$$

$\square$

Example 2.11. Next we integrate the exponential $e^{\lambda x}$. First we consider the lower limit $a = -\infty$,

$$_{-\infty}\mathcal{D}_x^{-\nu}e^{\lambda x} = \frac{1}{\Gamma(\nu)}\int_{-\infty}^x (x-x')^{\nu-1}e^{\lambda x'}dx'. \qquad (2.68)$$

Making the change of variable $t = x - x'$, we find

$$_{-\infty}\mathcal{D}_x^{-\nu}e^{\lambda x} = \frac{e^{\lambda x}}{\Gamma(\nu)}\int_\infty^0 t^{\nu-1}e^{-\lambda t}(-dt) = \lambda^{-\nu}e^{\lambda x}. \qquad (2.69)$$

Next we consider the lower limit $a = 0$, which yields

$$_0\mathcal{D}_x^{-\nu}e^{\lambda x} = \frac{1}{\Gamma(\nu)}\int_0^x (x-x')^{\nu-1}e^{\lambda x'}dx'$$

$$= x^\nu \sum_{n=0}^\infty \frac{(\lambda x)^n}{\Gamma(n+\nu+1)} = x^\nu E_{1,\nu+1}(\lambda x), \qquad (2.70)$$

where we have used Eq. (2.34). $\qquad\qquad\qquad\qquad\qquad\qquad \square$

From now on we use a short-hand notation, dropping the subscripts in the Riemann–Liouville integration, $_0\mathcal{D}_t^{-\nu} \equiv \mathcal{D}^{-\nu}$ for $a = 0$. Note that all properties of fractional integration are the same for both $a = 0$ and $a \neq 0$. Let us prove a "group property"

$$\mathcal{D}^{-\nu}\left[\mathcal{D}^{-\mu}f(t)\right] = \mathcal{D}^{-(\nu+\mu)}f(t) = \mathcal{D}^{-\mu}\left[\mathcal{D}^{-\nu}f(t)\right], \qquad (2.71)$$

which is based on Dirichlet's formula

$$\int_a^b dx \int_a^x G(x,y)dy = \int_a^b dy \int_y^b G(x,y)dx. \qquad (2.72)$$

Applying this expression to Eq. (2.71), we have

$$\frac{1}{\Gamma(\nu)\Gamma(\mu)}\int_0^t (t-\tau)^{\mu-1}d\tau \int_0^\tau (\tau-t')^{\nu-1}f(t')dt'$$

$$= \frac{1}{\Gamma(\nu)\Gamma(\mu)}\int_0^t f(t')dt' \int_{t'}^t (t-\tau)^{\mu-1}(\tau-t')^{\nu-1}d\tau$$

$$= \frac{1}{\Gamma(\nu)\Gamma(\mu)}\int_0^t (t-t')^{\mu+\nu-1}f(t')dt'\,B(\mu,\nu)dt'$$

$$= \frac{1}{\Gamma(\mu+\nu)}\int_0^t (t-t')^{\mu+\nu-1}f(t')dt' = \mathcal{D}^{-(\mu+\nu)}f(t). \qquad (2.73)$$

Example 2.12.

$$\mathcal{D}^{-\mu}t^\nu E_{1,\nu+1}(\lambda t) = \mathcal{D}^{-\mu}\left(\mathcal{D}^{-\nu}e^{\lambda t}\right) = \mathcal{D}^{-(\mu+\nu)}e^{\lambda t} = t^{\mu+\nu}E_{1,\mu+\nu+1}(\lambda t), \qquad (2.74)$$

where we have used Eq. (2.70) twice. $\qquad\qquad\qquad\qquad\qquad \square$

2.4.1 Properties of fractional integration with integer derivatives

To prepare for the definition of fractional derivatives, we present a few more properties of the fractional integration $\mathcal{D}^{-\nu} f(t)$. For a rigorous treatment see, e.g., Ref. [40]. First we consider a fractional integral $\mathcal{D}^{-\nu-1}$ of the derivative $\mathcal{D} \equiv d/dt$. Performing integration by part, we have

$$\mathcal{D}^{-\nu-1}\Big[\mathcal{D}f(t)\Big] = \frac{1}{\Gamma(\nu+1)} \int_0^t (t-\tau)^\nu \frac{df(\tau)}{d\tau} d\tau = \mathcal{D}^{-\nu} f(t) - \frac{f(0)}{\Gamma(\nu+1)} t^\nu.$$
(2.75)

Replacing ν by $\nu+1$ and $f(t)$ by $\mathcal{D}f$ in Eq. (2.75) yields

$$\mathcal{D}^{-\nu-2}\Big[\mathcal{D}^2 f(t)\Big] = \mathcal{D}^{-\nu-1}\Big[\mathcal{D}f(t)\Big] - \frac{\mathcal{D}f(0)}{\Gamma(\nu+2)} t^{\nu+1}$$

$$= \mathcal{D}^{-\nu} f(t) - \frac{f(0)}{\Gamma(\nu+1)} t^\nu - \frac{\mathcal{D}f(0)}{\Gamma(\nu+2)} t^{\nu+1}. \quad (2.76)$$

Repeating this procedure p times, we have

$$\mathcal{D}^{-\nu-p}\Big[\mathcal{D}^p f(t)\Big] = \mathcal{D}^{-\nu} f(t) - \sum_{k=0}^{p-1} \frac{t^{k+\nu}}{\Gamma(\nu+k+1)} \mathcal{D}^k f(0).$$
(2.77)

The second property concerns the differentiation of fractional integration $\mathcal{D}\Big[\mathcal{D}^{-\nu} f(t)\Big]$. We carry out the change of variable in the fractional integration $\tau = t - \zeta^{1/\nu}$, which yields

$$\mathcal{D}^{-\nu} f(t) = \frac{1}{\Gamma(\nu)} \int_0^t (t-\tau)^{\nu-1} f(\tau) d\tau = \frac{1}{\Gamma(\nu+1)} \int_0^{t^\nu} f\Big(t - \zeta^{1/\nu}\Big) d\zeta.$$
(2.78)

Then differentiating with respect to t, we obtain

$$\mathcal{D}\Big[\mathcal{D}^{-\nu} f(t)\Big] = \frac{f(0)}{\Gamma(\nu)} t^{\nu-1} + \frac{1}{\Gamma(\nu+1)} \int_0^{t^\nu} \frac{\partial}{\partial t} f(t - \zeta^{1/\nu}) d\zeta.$$
(2.79)

Now returning to τ by changing $t - \zeta^{1/\nu} = \tau$, we find

$$\mathcal{D}\Big[\mathcal{D}^{-\nu} f(t)\Big] = \mathcal{D}^{-\nu}\Big[\mathcal{D}f(t)\Big] + \frac{f(0)}{\Gamma(\nu)} t^{\nu-1}.$$
(2.80)

Repeating the differentiation, we perform the following chain of transformations,

$$\mathcal{D}^2\Big[\mathcal{D}^{-\nu} f(t)\Big] = \mathcal{D}\Big[\mathcal{D}^{-\nu}\Big[\mathcal{D}f(t)\Big]\Big] + \frac{f(0)}{\Gamma(\nu-1)} t^{\nu-2}$$

$$= \mathcal{D}^{-\nu}\left[\mathcal{D}^2 f(t)\right] + \frac{\mathcal{D}f(0)}{\Gamma(\nu)}t^{\nu-1} + \frac{f(0)}{\Gamma(\nu-1)}t^{\nu-2}. \quad (2.81)$$

Repeating this differentiation p times, we obtain

$$\mathcal{D}^p\left[\mathcal{D}^{-\nu} f(t)\right] = \mathcal{D}^{-\nu}\left[\mathcal{D}^p f(t)\right] + \sum_{k=0}^{p-1} \frac{t^{\nu+k-p}}{\Gamma(\nu+k-p+1)}\mathcal{D}^k f(0). \quad (2.82)$$

The final step is to show that

$$\mathcal{D}^p\left[\mathcal{D}^{-\nu} f(t)\right] = \mathcal{D}^{p-\nu} f(t) = \mathcal{D}^{-(\nu-p)} f(t) \quad (2.83)$$

for $\nu > p$ and p integer. To prove Eq. (2.83), let us consider the fractional integral $\mathcal{D}^{p-\nu-1} f(t)$, which exists since $\nu+1-p > 0$. Next, we differentiate this integral and use property (2.80), where we replace ν by $\nu - p + 1 > 0$. This yields

$$\mathcal{D}\left[\mathcal{D}^{p-\nu-1} f(t)\right] = \mathcal{D}^{p-\nu-1}\left[\mathcal{D}f(t)\right] + \frac{f(0)}{\Gamma(\nu-p+1)}t^{\nu-p}. \quad (2.84)$$

For the fractional integration in Eq. (2.84), we use property (2.75), where ν is replaced by $\nu - p$,

$$\mathcal{D}^{p-\nu-1}\left[\mathcal{D}f(t)\right] = \mathcal{D}^{p-\nu} f(t) - \frac{f(0)t^{\nu-p}}{\Gamma(\nu-p+1)}. \quad (2.85)$$

Substituting this into Eq. (2.84), we obtain

$$\mathcal{D}\left[\mathcal{D}^{p-\nu-1} f(t)\right] = \mathcal{D}^{p-\nu} f(t). \quad (2.86)$$

Differentiating this expression again, we find that $\mathcal{D}^2\left[\mathcal{D}^{p-\nu-2} f(t)\right] = \mathcal{D}^{p-\nu} f(t)$, and repeating this p times, we arrive at Eq. (2.83).

2.5   Fractional derivative

Formally, fractional differentiation is defined as the inverse operation of fractional integration:

$$\mathcal{D}^{-\nu} f(t) \Longleftrightarrow \mathcal{D}^\nu f(t), \quad \nu \longleftrightarrow -\nu. \quad (2.87)$$

However, the integral $\mathcal{D}^\nu f(t)$ diverges for $\nu > 0$:

$$\mathcal{D}^\nu f(t) = \frac{1}{\Gamma(-\nu)} \int_0^t (t-\tau)^{-\nu-1} f(\tau) d\tau. \quad (2.88)$$

Therefore, we must perform a regularization procedure of the integral (2.88) to obtain the correct definition of the fractional derivative. To this end, let us introduce an integer n such that $n - 1 < \nu < n$ and let us consider the well defined fractional integral $\mathcal{D}^{\nu-n} f(t)$ and then differentiate it n times. This yields

$$\mathcal{D}^{\nu} f(t) \Longrightarrow \mathcal{D}^n \left[\mathcal{D}^{-(n-\nu)} f(t) \right], \tag{2.89}$$

which is known as the Riemann–Liouville fractional derivative of order ν,

$$^{RL}\mathcal{D}^{\nu} f(t) = \frac{d^n}{dt^n} \frac{1}{\Gamma(n - \nu)} \int_0^t (t - \tau)^{n-\nu-1} f(\tau) d\tau. \tag{2.90}$$

In the general case, the lower limit can be arbitrary and the Riemann–Liouville derivative reads

$$^{RL}_a\mathcal{D}^{\nu}_x f(x) = \frac{d^n}{dx^n} \frac{1}{\Gamma(n - \nu)} \int_a^x (x - x')^{n-\nu-1} f(x') dx', \quad n - 1 < \nu < n. \tag{2.91}$$

The regularization procedure (2.89) or (2.90) is not unique. It also can be performed as

$$\mathcal{D}^{\nu} f(t) \Rightarrow \mathcal{D}^{-(n-\nu)} \left[\mathcal{D}^n f(t) \right], \tag{2.92}$$

which is known as a Caputo fractional derivative,

$$^C_a\mathcal{D}^{\nu}_x f(x) = \frac{1}{\Gamma(n - \nu)} \int_a^x (x - y)^{n-\nu-1} \frac{d^n}{dy^n} f(y) dy, \quad n - 1 < \nu < n. \tag{2.93}$$

The relation between the Caputo and the Riemann–Liouville fractional derivatives follows from Eq. (2.82), if we replace ν by $n - \nu$ and p by n,

$$^{RL}_a\mathcal{D}^{\nu}_x f(x) = {}^C_a\mathcal{D}^{\nu}_x f(x) + \sum_{k=0}^{n-1} \frac{x^{-\nu+k}}{\Gamma(k - \nu + 1)} f^{(k)}(a), \tag{2.94}$$

where $f^{(k)}(a) \equiv d^k f(x)/dx^k|_{x=a}$.

Example 2.13. Let us fractionally differentiate a power law function $f(t) = t^{\mu}$ with $\mu > -1$. The Riemann–Liouville fractional derivative is given by

$$
\begin{aligned}
^{RL}\mathcal{D}^{\nu} t^{\mu} &= \frac{d^n}{dt^n} \frac{1}{\Gamma(n - \nu)} \int_0^t (t - \tau)^{n-\nu-1} t^{\mu} d\tau \\
&= \frac{1}{\Gamma(n - \nu)} \mathcal{D}^n t^{n-\nu+\mu} \int_0^1 (1 - \tau)^{n-\nu-1} \tau^{\mu} = \frac{\Gamma(\mu + 1)}{\Gamma(\mu - \nu + 1)} t^{\mu-\nu}.
\end{aligned}
\tag{2.95}
$$

For $\mu = 0$, we find

$$^{RL}\mathcal{D}^{\nu}[1] = \frac{t^{-\nu}}{\Gamma(1-\nu)}. \tag{2.96}$$

The Caputo fractional derivative yields

$$^{C}\mathcal{D}^{\nu}[1] = 0. \tag{2.97}$$

$\square$

Throughout the rest of this book we will use short-hand notations for the case where the lower limit is $a = 0$,

$$^{RL}_{0}\mathcal{D}^{\nu}_{t} \equiv \mathcal{D}^{\nu}, \tag{2.98a}$$

$$^{C}_{0}\mathcal{D}^{\nu}_{t} \equiv \frac{\partial^{\nu}}{\partial t^{\nu}} \equiv \partial^{\nu}. \tag{2.98b}$$

We discuss two important properties of fractional derivatives. First, the group property (2.71), which holds for fractional integration, is violated,

$$\mathcal{D}^{\mu}\mathcal{D}^{\nu}f(t) \neq \mathcal{D}^{\mu+\nu}f(t). \tag{2.99}$$

Example 2.14. To prove property (2.99), we consider an example from Ref. [40]. Let $f(t) = t^{1/2}$ and $\mu = 1/2$, while $\nu = 3/2$. Then

$$\mathcal{D}^{\mu}t^{1/2} = \frac{1}{2}\sqrt{\pi} \quad \text{and} \quad \mathcal{D}^{\nu}t^{1/2} = \frac{\Gamma(3/2)}{\Gamma(0)}t^{-1} = 0. \tag{2.100}$$

Therefore, $\mathcal{D}^{\mu}\left[\mathcal{D}^{\nu}t^{\frac{1}{2}}\right] = 0$, while

$$\mathcal{D}^{\nu}\left[\mathcal{D}^{\mu}t^{\frac{1}{2}}\right] = \frac{1}{2}\sqrt{\pi}\mathcal{D}^{\frac{3}{2}}[1] = -\frac{1}{4}t^{-\frac{3}{2}} \tag{2.101}$$

and

$$\mathcal{D}^{\mu+\nu}t^{\frac{1}{2}} = \frac{d^{2}}{dt^{2}}t^{\frac{1}{2}} = -\frac{1}{4}t^{-\frac{3}{2}}. \tag{2.102}$$

This example proves Eq. (2.99). $\square$

The second property is Leibniz's formula for fractional derivatives, which reads for the product of two functions $f(t)g(t)$:

$$\mathcal{D}^{\nu}\left[f(t)g(t)\right] = \sum_{k=0}^{\infty} \binom{\nu}{k}\left[\mathcal{D}^{k}g(t)\right]\left[\mathcal{D}^{\nu-k}f(t)\right]. \tag{2.103}$$

This expression is convenient if $g(t)$ is a polynomial. For example for $g(t) = t$, Eq. (2.103) yields

$$\mathcal{D}^{\nu}\left[f(t)t\right] = t\mathcal{D}^{\nu}f(t) + \nu\mathcal{D}^{\nu-1}f(t). \tag{2.104}$$

2.6 Laplace transform

First, we consider the Laplace transform of a fractional integral. The Laplace transform of a function $f(t)$ is given by

$$\hat{f}(s) = \mathcal{L}[f(t)](s) = \int_0^\infty e^{-st} f(t)dt, \tag{2.105}$$

and the inverse Laplace transformation is

$$f(t) = \mathcal{L}^{-1}\hat{f}(s) = \frac{1}{2\pi i} \int_{c-i\infty}^{c+i\infty} e^{st} \hat{f}(s)ds. \tag{2.106}$$

One of the most important properties is the Laplace transform of a convolution integral

$$\mathcal{L}[f(t) * g(t)] = \mathcal{L}\left[\int_0^t f(t-\tau)g(\tau)d\tau\right] = \hat{f}(s)\hat{g}(s), \tag{2.107}$$

which makes it extremely efficient and pivotal for fractional integration together with the following standard results,

$$\mathcal{L}[t^\mu] = \Gamma(\mu+1)s^{-\mu-1}, \quad \mu > -1, \tag{2.108a}$$

$$\mathcal{L}[e^{at}] = (s-a)^{-1}, \tag{2.108b}$$

$$\mathcal{L}[e^{at}t^{\mu-1}] = \Gamma(\mu)(s-a)^{-\mu}, \quad \mu > 0. \tag{2.108c}$$

Applying these properties to the fractional integral

$$\mathcal{D}^{-\nu}f(t) = \frac{1}{\Gamma(\nu)} \int_0^t (t-\tau)^{\nu-1} f(\tau)d\tau, \tag{2.109}$$

we obtain

$$\mathcal{L}[\mathcal{D}^{-\nu}f(t)] = \frac{1}{\Gamma(\nu)} \mathcal{L}[t^{\nu-1}] \mathcal{L}[f(t)] = s^{-\nu}\hat{f}(s), \quad \nu > 0. \tag{2.110}$$

It is important to note here that the limit of $\nu \to 0$ and the Laplace transform do not commute [40],

$$\lim_{\nu \to 0} \frac{t^{\nu-1}}{\Gamma(\nu)} = 0, \quad \text{while} \quad \lim_{\nu \to 0} \mathcal{L}\left[\frac{t^{\nu-1}}{\Gamma(\nu)}\right] = 1. \tag{2.111}$$

Example 2.15. Let us evaluate the Laplace transforms of two fractional integrals using Eqs. (2.108):

$$\mathcal{L}[\mathcal{D}^{-\nu}t^\mu] = \Gamma(\mu+1)s^{-(\mu+\nu+1)}, \quad \nu > 0, \quad \mu > -1, \tag{2.112}$$

$$\mathcal{L}[\mathcal{D}^{-\nu}e^{at}] = s^{-\nu}(s-a)^{-1}, \quad \nu > 0. \tag{2.113}$$

$\square$

Let us calculate the Laplace transforms of the fractional integral of the derivative and the derivative of the fractional integral [40]. Equation (2.110) implies

$$\mathcal{L}\left[\mathcal{D}^{-\nu}\left[\mathcal{D}f(t)\right]\right] = s^{-\nu}\mathcal{L}\left[\mathcal{D}f(t)\right] = s^{1-\nu}\hat{f}(s) - f(0)s^{-\nu}. \qquad (2.114)$$

Employing Eqs. (2.80), (2.114), and (2.108), we find

$$\mathcal{L}\left[\mathcal{D}\left[\mathcal{D}^{-\nu}f(t)\right]\right] = \mathcal{L}\left[\mathcal{D}^{-\nu}\left[\mathcal{D}f(t)\right]\right] + \frac{f(0)}{\Gamma(\nu)}\mathcal{L}\left[t^{\nu-1}\right] = s^{1-\nu}\hat{f}(s). \quad (2.115)$$

When $0 < \nu < 1$, Eqs. (2.114) and (2.115) correspond to the Laplace transform of the Caputo and Riemann–Liouville fractional derivatives, respectively.

We generalize these calculations to arbitrary ν and consider the Laplace transform of fractional derivatives D^ν and ∂^ν, see Eqs. (2.98a) and (2.98b). For the Laplace transform of the Caputo fractional derivative we obtain

$$\mathcal{L}\left[\partial^\nu f(t)\right] = \mathcal{L}\left[\mathcal{D}^{-(n-\nu)}\left[\mathcal{D}^n f(t)\right]\right] = s^{n-\nu}\mathcal{L}\left[\mathcal{D}^n f(t)\right]$$

$$= s^\nu \hat{f}(s) - \sum_{k=0}^{n-1} f^{(k)}(0)s^{\nu-k-1}. \qquad (2.116)$$

For the Riemann–Liouville fractional derivative, the Laplace transform reads

$$\mathcal{L}\left[D^\nu f(t)\right] = \mathcal{L}\left[\mathcal{D}^n\left[\mathcal{D}^{-(n-\nu)}f(t)\right]\right] = s^n\mathcal{L}\left[\mathcal{D}^{-(n-\nu)}f(t)\right]$$

$$- \sum_{k=0}^{n-1} s^{n-k-1}\mathcal{D}^k\left[\mathcal{D}^{-(n-\nu)}f(t)\right]\Big|_{t=0}$$

$$= s^\nu \hat{f}(s) - \sum_{k=0}^{n-1} s^{n-k-1}\mathcal{D}^{k-n+\nu}f(0). \qquad (2.117)$$

To Laplace transform the Caputo fractional derivative, we need to know the value of $f(t)$ at the lower limit of integration, $t = 0$, and its $n-1$ *integer derivatives* at $t = 0$, $f^k(0)$. The situation is quite different for the Laplace transform of the Rieman-Liouville fractional derivative; here we need to know all n *fractional derivatives* at $t = 0$.

Example 2.16. The Mittag–Leffler function is a solution of a Caputo fractional differential equation. Let us consider the Cauchy problem

$$\partial^\nu f(t) = -\lambda f(t), \quad 0 < \nu < 1, \qquad (2.118)$$

with the initial condition $f_0 = f(t = 0)$. Laplace transforming this equation, we obtain

$$s^\nu \hat{f} - s^{\nu-1} f_0 = -\lambda \hat{f}, \tag{2.119}$$

which yields, using Eq. (2.34) with $\alpha = \nu$ and $\beta = 1$,

$$f(t) = \frac{f_0}{2\pi i} \int_{c-i\infty}^{c+i\infty} \frac{s^{\nu-1} e^{st} ds}{s^\nu + \lambda} = f_0 E_\nu\left(-\lambda t^\nu\right). \tag{2.120}$$

$\square$

2.7 Weyl and Riesz fractional derivatives

As discussed above, there is no constraint on the lower limit a. For example, when $a = -\infty$, the resulting fractional derivative is the Weyl derivative,

$$\mathcal{W}^\nu f(x) \equiv {}_{-\infty}\mathcal{W}_x^\nu f(x) \equiv {}_{-\infty}^W\mathcal{D}_x^\nu f(x) = \frac{1}{\Gamma(-\nu)} \int_{-\infty}^x \frac{f(x')dx'}{(x-x')^{1+\nu}}. \tag{2.121}$$

If we impose the physically reasonable condition $f(-\infty) = 0$ together with its n derivatives, where $n - 1 < \nu < n$, then

$${}_{-\infty}^W\mathcal{D}_x^\nu f(x) = {}_{-\infty}^{RL}\mathcal{D}_x^\nu f(x) = {}_{-\infty}^C\mathcal{D}_x^\nu f(x). \tag{2.122}$$

One also has $\mathcal{W}^\nu e^{\lambda x} = \lambda^\nu e^{\lambda x}$. This property is convenient for the Fourier transform $\mathcal{F}[f(x)](k) = \tilde{f}(k)$, which yields

$$\mathcal{F}\left[\mathcal{W}^\nu f(x)\right](k) = (-ik)^\nu \tilde{f}(k). \tag{2.123}$$

Let us prove this property, following a result of Ref. [41], based on Dirichlet's test for uniform convergence.

The Fourier transformation of a function $h(x)$ can be defined on $x \in (-\infty, \infty)$, such that the Fourier transform $\tilde{h}(k)$ and the original function $h(x)$ are related by the Fourier integral pair,

$$\tilde{h}(k) = \mathcal{F}[h(x)](k) = \int_{-\infty}^\infty h(x)e^{ikx} dx, \tag{2.124a}$$

$$h(x) = \frac{1}{2\pi} \int_{-\infty}^\infty \tilde{h}(k)e^{-ikx} dk. \tag{2.124b}$$

An important property is the Fourier transform of a convolution integral,

$$\mathcal{F}[h(x) * f(x)] = \mathcal{F}\left[\int_{-\infty}^\infty h(x-x')f(x')dx'\right] = \tilde{h}(k)\tilde{f}(k). \tag{2.125}$$

The Laplace transform of $h(x) = t^{\mu-1}/\Gamma(\mu)$ immediately follows from Eq. (2.1), which yields

$$\hat{h}(\zeta) = \frac{1}{\Gamma(\mu)} \int_0^\infty x^{\mu-1} e^{-\zeta x} dx = \zeta^{-\mu}. \tag{2.126}$$

Let us take $\zeta = -ik$ with k real. Then for $0 < \mu < 1$, the integral (2.126) satisfies Dirichlet's criterion of uniform convergence[3] [47], and the Fourier transform of the function $h_+(x)$ exists, where $x \in (-\infty, \infty)$ and $h_+(x) = h(x)$ for $x > 0$ and $h_+(x) = 0$ for $x \le 0$, and it reads $\hat{h}_+(k) = (-ik)^\mu$. Therefore, the Fourier transform of the fractional integral yields

$$\mathcal{F}\left[{}_{-\infty}\mathcal{D}_x^{-\nu}f(x)\right] = \mathcal{F}\left[h_+(x)\right]\mathcal{F}\left[f(x)\right] = (-ik)^{-\nu}\tilde{f}(k), \qquad 0 < \nu < 1.$$
$$(2.127)$$

Let us represent the Weyl fractional derivative in the Caputo form for arbitrary $\nu > 0$, taking an integer n as $n - 1 < \nu < n$. This yields

$$\substack{W\\-\infty}\mathcal{D}_x^\nu f(x) = \substack{C\\-\infty}\mathcal{D}_x^\nu f(x) = {}_{-\infty}\mathcal{D}_x^{-(n-\nu)}\left[f^{(n)}(x)\right]. \qquad (2.128)$$

Therefore, the Fourier transform of the Weyl fractional derivative yields the result of Eq. (2.123):

$$\mathcal{F}\left[\substack{W\\-\infty}\mathcal{D}_x^\nu f(x)\right] = \mathcal{F}\left[{}_{-\infty}\mathcal{D}_x^{-(n-\nu)}\left[f^{(n)}(x)\right]\right]$$
$$= (-ik)^{\nu-n}\mathcal{F}\left[f^{(n)}(x)\right] = (-ik)^\nu\tilde{f}(k), \qquad (2.129)$$

where we impose the zero condition for the l-th derivative $f^{(l)}(x = -\infty) = 0$, and $l = 0, 1, 2, \ldots, n - 1$.

The fractional derivation with the fixed lower limit is also called the left fractional derivative. One can introduce the right fractional derivative, where the upper limit a is fixed and $a > x$. For example, the formal definition of the right Weyl fractional derivative is

$$_x\mathcal{W}_\infty^\nu f(x) \equiv {}^W_x\mathcal{D}_\infty^\nu f(x) = \frac{1}{\Gamma(-\nu)}\int_x^\infty \frac{f(y)dy}{(y-x)^{-(1+\nu)}}, \qquad (2.130)$$

which is regularized in either Riemann–Liouville (2.90) or Caputo (2.93) forms. This derivative, acting on $e^{-\lambda x}$ for $\lambda > 0$, yields $_x\mathcal{W}_\infty^\nu e^{-\lambda x} = \lambda^\nu e^{-\lambda x}$.

Example 2.17. Let us consider the Weyl derivative $_{-\infty}\mathcal{W}_{-x}^\nu f(-x)$ at $-x$, which yields the following chain of transformations

$$_{-\infty}\mathcal{W}_{-x}^\nu f(-x) = \frac{1}{\Gamma(-\nu)}\int_{-\infty}^{-x}(-x-x')^{-\nu-1}f(-x')dx' \quad \underset{x' \to -x'}{=}$$
$$= \frac{1}{\Gamma(-\nu)}\int_\infty^x (x'-x)^{-\nu-1}f(x')(-dx')$$

---

[3] Then integral over any finite interval $x \in [a, b]$ is bounded: $\left|\int_a^b e^{ikx}dx\right| < 1/|k|$, while for $0 < \mu < 1$, the function $x^{\mu-1}$ decreases monotonically to zero. These conditions correspond to Dirichlet's criterion of uniform convergence [47].

$$= \frac{1}{\Gamma(-\nu)} \int_x^\infty (x'-x)^{-\nu-1} f(x')dx' = {}_x W_\infty^\nu f(x).$$

(2.131)

Performing the Fourier transform in the framework of the Caputo regularization (2.129), we obtain a chain of calculations,

$$\mathcal{F}\left[{}_x^W \mathcal{D}_\infty^\nu f(x)\right]$$

$$= \mathcal{F}\left[{}_x \mathcal{D}_\infty^{-(n-\nu)}\left[f_{(-x)}^{(n)}(x)\right]\right]$$

$$= \mathcal{F}\left[\frac{1}{\Gamma(n-\nu)} \int_x^\infty (x'-x)^{n-\nu-1}\left(-\frac{d}{dx'}\right)^n f(x')dx'\right]$$

$$= \mathcal{F}\left[\frac{1}{\Gamma(n-\nu)}\frac{1}{2\pi} \int_{-\infty}^\infty d\kappa (i\kappa)^n \tilde{f}(\kappa) \int_x^\infty (x'-x)^{n-\nu-1} e^{-i\kappa x'} dx'\right]$$

$$= \frac{1}{2\pi} \int_{-\infty}^\infty d\kappa (i\kappa)^n \tilde{f}(\kappa)\mathcal{F}\left[e^{-i\kappa x}\right]\frac{1}{\Gamma(n-\nu)} \int_0^\infty y^{n-\nu-1} e^{-i\kappa y} dy$$

$$= (ik)^n \tilde{f}(k)\frac{1}{\Gamma(n-\nu)} \int_0^\infty y^{n-\nu-1} e^{-iky} dy.$$

(2.132)

Substitution $ik = s$ in the last integration yields $s^{-(n-\nu)}\Gamma(n-\nu)$, and the final result is

$$\mathcal{F}\left[{}_x^W \mathcal{D}_\infty^\nu f(x)\right] = (ik)^\nu \tilde{f}(k).$$

(2.133)

□

We will also use the so-called Riesz fractional derivative and introduce it by noting that

$$\frac{1}{2\Gamma(\nu)\cos\frac{\nu\pi}{2}} \int_a^b \frac{f(y)dy}{|x-y|^{1-\nu}}$$

(2.134)

is the Riesz fractional integral on the finite interval $[a,b]$ [42] with $a \le x \le b$ and $0 < \nu < 1$. It can be represented as the sum of the left and right Riemann–Liouville fractional integrals

$$\int_a^x \frac{f(y)dy}{(x-y)^{1-\nu}} + \int_x^b \frac{f(y)dy}{(y-x)^{1-\nu}}.$$

(2.135)

Consequently, the fractional Riesz derivative $(-\Delta)^{\frac{\nu}{2}}$ on the entire x-axis can also be represented by the Weyl derivatives

$$(-\Delta)^{\frac{\nu}{2}} \equiv {}_\infty \mathcal{D}_{|x|}^\nu f(x) = \frac{1}{2\cos\frac{\nu\pi}{2}}\left[{}_{-\infty}^W \mathcal{D}_x^\nu f(x) + {}_{-\infty}^W \mathcal{D}_{-x}^\nu f(-x)\right].$$

(2.136)

Example 2.18. To conclude this chapter, we consider two examples involving the Riesz derivative. The first one concerns the Riesz derivative of order ν on a finite interval $x \in [-L, L]$ with $1 < \nu < 2$. We write it in terms of the inverse Fourier transform,

$$_L\mathcal{D}^\nu_{|x|}f(x) = \frac{1}{2\pi}\int_{-L}^{L}dx'\int_{-\infty}^{\infty}|k|^\nu e^{-ik(x-x')}f(x')dk. \qquad (2.137)$$

The inverse Fourier transformation in Eq. (2.137) as an integration with respect to k yields

$$\frac{1}{2\pi}\int_{-\infty}^{\infty}|k|^\nu e^{-ik(x-x')}dk = \frac{1}{\pi}\int_{0}^{\infty}k^\nu\cos\big[k(x-x')\big]dk$$

$$= \frac{1}{2\pi}\Big(-\frac{d^2}{dx^2}\Big)\int_{0}^{\infty}k^{\nu-2}\cos\big[k(x-x')\big]$$

$$= \Big(-\frac{d^2}{dx^2}\Big)\frac{(x-x')^{1-\nu}}{\Gamma(2-\nu)\cos\big[\frac{(2-\nu)\pi}{2}\big]}$$

$$= \frac{(x-x')^{-\nu-1}}{\Gamma(-\nu)\cos\big(\frac{\nu\pi}{2}\big)}. \qquad (2.138)$$

We have taken into account that $x - x' > 0$. In the opposite case of $x - x' < 0$, the term in the numerator in Eq. (2.138) reads $(x' - x)^{-\nu-1}$, and the result of the integration with respect to k is

$$\frac{|x-x'|^{-\nu-1}}{\Gamma(-\nu)\cos\big(\frac{\nu\pi}{2}\big)}. \qquad (2.139)$$

Substituting this result into Eq. (2.137), we obtain eventually

$$_L\mathcal{D}^\nu_{|x|}f(x) = \frac{1}{\Gamma(-\nu)\cos\big(\frac{\nu\pi}{2}\big)}\int_{-L}^{L}|x-x'|^{-\nu-1}f(x')dx'. \qquad (2.140)$$

$\square$

Example 2.19. Alternatively, we can introduce the Riesz fractional derivative by means of the inverse Fourier transformation with a slightly different Fourier transform, namely $-|k|^{-\mu}$,

$$\frac{\partial^\mu f(x)}{\partial|x|^\mu} = \mathcal{F}^{-1}\Big[-|k|^{-\mu}f(k)\Big]. \qquad (2.141)$$

Let us consider the fractional in both space and time Fokker–Planck equation,

$$^C\mathcal{D}^\alpha_t f(x,t) = D\frac{\partial^\mu f(x,t)}{\partial|x|^\mu}, \qquad (2.142)$$

where Eq. (2.142) is a general form of the FFPE, with $0 < \alpha < 1$ and $0 < \mu < 2$ and a generalized diffusion coefficient D. In Fourier-Laplace space, it reads

$$s^\alpha \hat{\tilde{f}}(k, s) = -D|k|^\mu \hat{\tilde{f}}(k, s) + s^{-\alpha}. \tag{2.143}$$

This equation can be solved in terms of Fox H-functions. Specifically, we obtain from Eq. (2.143) the Mittag–Leffler function, a particular case of the Fox H function, see (2.44), namely

$$\tilde{f}(k, t) = \mathcal{L}^{-1} \left[\frac{s^{\alpha - 1}}{s^\alpha + D|k|^\mu} \right] = E_\alpha \left(-Dt^\alpha |k|^\mu \right). \tag{2.144}$$

Let us establish this relation in an alternative way by writing the Mittag–Leffler function in the form of a Mellin transform $\mathcal{M}[f(z)](p) = f(p)$. Using the well known formula for the Mellin transform [48],

$$\mathcal{M}[f(z)](p) = \frac{1}{\Gamma(1-p)} \mathcal{M}\left[\mathcal{L}[f(z)](s) \right](1-p), \tag{2.145}$$

we represent the Mittag–Leffler function as the inverse Mellin transform via a chain of transformations

$$\begin{aligned}
E_\alpha(-z^\alpha) &= \mathcal{M}^{-1} \left\{ \frac{1}{\Gamma(1-p)} \mathcal{M}\left[\mathcal{L}[E_\alpha(-z^\alpha)](s) \right](1-p) \right\} \\
&= \mathcal{M}^{-1} \left\{ \frac{1}{\Gamma(1-p)} \left[\int_0^\infty \frac{s^{\alpha-1} s^{p-1} ds}{s^\alpha + 1} \right](1-p) \right\} \\
&= \frac{1}{2\pi i} \int_C \frac{\Gamma(1 - p/\alpha)\Gamma(p/\alpha)}{\alpha \Gamma(1-p)} z^{-p} dp \\
&= H_{2,1}^{1,1} \left[z^\alpha \left| \begin{matrix} (0, 1) \\ (0, 1), (1, \alpha) \end{matrix} \right. \right].
\end{aligned} \tag{2.146}$$

Here $z^\alpha = Dt^\alpha |k|^\mu$. To perform the inverse Fourier transformation, which reduces to the Fourier-cosine transformation, we employ the Mellin transform [48] in combination with the cosine Fourier transform $\frac{1}{\pi}\mathcal{F}_c$,

$$\mathcal{M}\left[\mathcal{F}_c[\phi(k)](x) \right](p) = \Gamma(p) \cos\left(\frac{p\pi}{2} \right) \mathcal{M}[\phi(k)](1-p). \tag{2.147}$$

We also use the following property of the Mellin transform [48],

$$\mathcal{M}\left[z^\rho \phi(az^\mu) \right](p) = \frac{1}{\mu} a^{-\frac{p+\rho}{\mu}} \mathcal{M}\left[\phi(z) \right] \left(\frac{p+\rho}{\mu} \right). \tag{2.148}$$

Taking these properties into account, we obtain that the Mellin transform of the Fox function in Eq. (2.146) is given by

$$\left[(Dt^\alpha)^{-\frac{p}{\mu}} \frac{\Gamma(1-p/\mu)\Gamma(p/\mu)}{\Gamma(1-p\alpha/\mu)}\right] (p \to 1-p)$$

$$= (Dt^\alpha)^{-\frac{1-p}{\mu}} \frac{\Gamma\left(\frac{1}{\mu}-\frac{p}{\mu}\right)\Gamma\left(1-\frac{1}{\mu}-\frac{p}{\mu}\right)}{\Gamma\left(1-\frac{\alpha}{\mu}+p\frac{\alpha}{\mu}\right)}. \tag{2.149}$$

The Fourier inversion of the Fox function in Eq. (2.146) reads,

$$f(x,t) = \mathcal{F}^{-1}[\tilde{f}(k,t)]$$

$$= \frac{1}{\mu\pi}(Dt^\alpha)^{-\frac{1}{\mu}}\mathcal{M}^{-1}$$

$$\times \left\{(Dt^\alpha)^{\frac{p}{\mu}}\Gamma(p)\cos\left(\frac{p\pi}{2}\right) \frac{\Gamma\left(\frac{1}{\mu}-\frac{p}{\mu}\right)\Gamma\left(1-\frac{1}{\mu}-\frac{p}{\mu}\right)}{\Gamma\left(1-\frac{\alpha}{\mu}+p\frac{\alpha}{\mu}\right)}\right\}$$

$$= \frac{1}{\mu}(Dt^\alpha)^{-\frac{1}{\mu}}\mathcal{M}^{-1}\left\{(Dt^\alpha)^{\frac{p}{\mu}} \frac{\Gamma(p)\Gamma\left(\frac{1}{\mu}-\frac{p}{\mu}\right)\Gamma\left(1-\frac{1}{\mu}-\frac{p}{\mu}\right)}{\Gamma\left(\frac{1}{2}+\frac{p}{2}\right)\Gamma\left(\frac{1}{2}-\frac{p}{2}\right)\Gamma\left(1-\frac{\alpha}{\mu}+p\frac{\alpha}{\mu}\right)}\right\}$$

$$= \frac{1}{\mu}(Dt^\alpha)^{-\frac{1}{\mu}} H_{2,1}^{3,3}\left[\frac{x}{(Dt^\alpha)^{\frac{1}{\mu}}} \middle| \begin{array}{l}(1-\frac{1}{\mu},\frac{1}{\mu}),(\frac{1}{2},\frac{1}{2}),(1-\frac{\alpha}{\mu},\frac{\alpha}{\mu})\\(0,1),(1-\frac{1}{\mu},\frac{1}{\mu}),(\frac{1}{2},\frac{1}{2})\end{array}\right]$$

$$= \frac{1}{\mu}|x|^{-1} H_{2,1}^{3,3}\left[\frac{|x|}{(Dt^\alpha)^{\frac{1}{\mu}}} \middle| \begin{array}{l}(1,\frac{1}{\mu}),(1,\frac{1}{2}),(1,\frac{\alpha}{\mu})\\(1,1),(1,\frac{1}{\mu}),(1,\frac{1}{2})\end{array}\right]$$

$$= \frac{1}{|x|} H_{2,1}^{3,3}\left[\frac{|x|^\mu}{(Dt^\alpha)} \middle| \begin{array}{l}(1,1),(1,\frac{\mu}{2}),(1,\alpha)\\(1,\mu),(1,1),(1,\frac{\mu}{2})\end{array}\right]. \tag{2.150}$$

Equation (2.150) shows an alternative way of the Mellin-cosine transform in Eq. (2.54), as well as some different representations of the same Fox H-function [44, 45].   □

Chapter 3

Markov processes

3.1 Stochastic processes

A Markov process, taking place separately along both the backbone and the branches of a comb, gives rise to the phenomenon of anomalous diffusion by virtue of the comb geometry. Markov processes are an important class of stochastic processes. To describe Markov processes, or present at least a few of their specific properties, we need to introduce first the concept of a stochastic process itself.

We present some basic material on Markov processes, following the concise introduction to the topic in Ref. [49]. We assume an elementary knowledge of probability theory. The literature on this subject is vast, and the reader has an abundance of books to choose from. Besides the fundamental work by Feller [50] and [51], an extended introduction to stochastic theory can be found, for example, in Refs. [52–55].

A stochastic process is a set of random variables, say $X(t)$, indexed by the variable t. The latter usually denotes time, with $t \geq 0$, and can belong to the set of natural or integer numbers or to the set of real numbers.[1] The family of random variables $X(t)$ is called a discrete-time stochastic process in the former case, and a continuous-time stochastic process in the latter case. Since the variable $X(t)$ takes on random values as a function of t, a stochastic process can be thought of a family of functions of time. The elements of this family are called realizations of the random process or sample paths. For example, at $t = t_1, t_2, \ldots, t_n$, we have the realization $X(t_1) = x_1, X(t_2) = x_2, \ldots, X(t_n) = x_n$. A basic piece of information for

---

[1]One uses the terms "stochastic" or "random," but not "chaotic." The latter term is applied to the unpredictable dynamics in deterministic nonlinear systems, which arises due to the dynamical local instability of trajectories, see, e.g., [56].

the stochastic process is the probability density function (pdf) $P(x,t)$ of $X(t)$ for each t. Namely,

$$P(x_j, t_j)dx_j = \Pr(x_j < X(t_j) < x_j + dx_j) \tag{3.1}$$

is the probability that $X(t_j) \in (x_j, x_j + dx)$ at $t = t_j$, where $j = 1, 2, \ldots, n$ and each x_j varies over its own region or field of realizations, $x_j \in \Omega_j$. Typically, the range of possible values of the stochastic process is time-independent, and $\Omega_j = \Omega$ for all j. More information about the process is contained in the joint pdf $P(x_2, t_2; x_1, t_1)dx_1dx_2$, which represents the probability that $x_1 < X(t_1) < x_1 + dx_1$ for $t = t_1$ and $x_2 < X(t_2) < x_2 + dx_2$ for $t = t_2$. Continuing in this way, we obtain a hierarchy of finite-dimensional joint pdfs $P(x_n, t_n; \ldots ; x_2, t_2; x_1, t_1)$. All pdfs must fulfill the normalization condition

$$\int_{\Omega_1} \cdots \int_{\Omega_n} P(x_n, t_n; \ldots ; x_2, t_2; x_1, t_1)dx_1 \cdots dx_n = 1, \tag{3.2}$$

the symmetry condition that $P(x_n, t_n; \ldots ; x_2, t_2; x_1, t_1)$ is invariant with respect to any permutation of the numbers $1, 2, \ldots, n$, and the compatability condition

$$\int_{\Omega_i} P(x_{n+1}, t_{n+1}; \ldots ; x_{i+1}, t_{i+1}; x_i, t_i; x_{i-1}, t_{i-1}; \ldots ; x_1, t_1)dx_i$$
$$= P(x_{n+1}, t_{n+1}; \ldots ; x_{i+1}, t_{i+1}; x_{i-1}, t_{i-1}; \ldots ; x_1, t_1). \tag{3.3}$$

In addition, the conditional probability density that $X(t_i) = x_i$ for $j + 1 \leq i \leq n$ given that $X(t_i) = x_i$ for $1 \leq i \leq j$, reads

$$P(x_n, t_n; \ldots ; x_{j+1}, t_{j+1} \mid x_j, t_j; \ldots; x_1, t_1) = \frac{P(x_n t_n; \ldots ; x_1, t_1)}{P(x_j, t_j; \ldots ; x_1, t_1)}. \tag{3.4}$$

This implies that

$$P(x_n, t_n; \ldots; x_1, t_1)$$
$$= P(x_n, t_n; \ldots; x_{j+1}, t_{j+1} \mid x_j, t_j; \ldots; x_1, t_1)P(x_j, t_j; \ldots; x_1, t_1). \tag{3.5}$$

This expression allows us to connect the joint pdf for the set $\{t_1, \ldots, t_j\}$ to the one for the set $\{t_1, \ldots, t_n\}$. Equations (3.3) and (3.5) imply that

$$P(x_2, t_2) = \int_{\Omega_1} P(x_2, t_2 \mid x_1, t_1)P(x_1, t_1)dx_1, \tag{3.6}$$

i.e., the conditional pdf plays the role of the *transition pdf*, or the Green's function, for the evolution of the pdf.

3.1.1 *Averaged characteristics of stochastic processes*

The hierarchy of finite-dimensional joint pdfs $P(x_n, t_n; \ldots; x_2, t_2; x_1, t_1)$ characterizes the random process. Frequently we do not need the full information contained in the joint pdfs, and average quantities or moments suffice. The moment of order n of the random variable $X(t)$ is defined as

$$\langle X^n(t) \rangle = \int_{\Omega} x^n P(x, t) dx. \tag{3.7}$$

The first moment $\langle X \rangle$ is also called the mean or expectation value. The combination of the first and the second moments

$$\sigma^2 = \langle (X - \langle X \rangle)^2 \rangle = \langle X^2 \rangle - \langle X \rangle^2 \tag{3.8}$$

is known as the variance or dispersion. Since our focus is on transport processes, and the random variable $X(t)$ typically represents the position of a particle, we call the second central moment σ^2 the mean squared displacement (MSD).

The n-th moment can be obtained from a generating function, which is the Fourier transform of the pdf

$$\langle e^{ikX(t)} \rangle = \int_{\Omega} e^{ikx} P(x, t) dx = \tilde{P}(k, t), \tag{3.9}$$

which is known as the *characteristic function* of the variable $X(t)$. Differentiating the characteristic function n times with respect to k at $k = 0$,

$$\left. \frac{d^n}{d\,k^n} \tilde{P}(k, t) \right|_{k=0}, \tag{3.10}$$

we obtain Eq. (3.7).

The covariance of two random variables, $X(t_1)$ and $X(t_2)$, often called the correlation function in the case of random processes, is defined as

$$\mathrm{Cor}\left(X(t_1), X(t_2)\right) = \langle [X(t_1) - \langle X(t_1) \rangle][X(t_2) - \langle X(t_2) \rangle] \rangle. \tag{3.11}$$

If the correlation functions vanishes,

$$\mathrm{Cor}\left(X(t_1), X(t_2)\right) = 0, \tag{3.12}$$

the two random variables are said to be uncorrelated. A stronger property is for the two variables to be independent,

$$P(x_1, t_1; x_2, t_2) = P(x_1, t_1) \cdot P(x_2, t_2), \tag{3.13}$$

which implies that the two variables are uncorrelated:

$$\mathrm{Cor}\left(X(t_1), X(t_2)\right) = \langle X(t_1) - \langle X(t_1) \rangle \rangle \cdot \langle X(t_2) - \langle X(t_2) \rangle \rangle = 0. \tag{3.14}$$

The converse does not hold; uncorrelated does not imply independent. For a random process with independent values at every instant of time, $X(t_1), \ldots, X(t_n)$, the joint pdf is

$$P(x_n, t_n; \ldots; x_1, t_1) = \prod_{i=1}^{n} P(x_i, t_i), \tag{3.15}$$

and such a process is called a purely or completely random process.

3.2   Markov processes

The dependence relation between the random variables $X(t_1), \ldots, X(t_n)$ characterizes the random process. In the preceding section we encountered the simplest type of dependence, namely none at all, which defines the class of purely or completely random processes. Clearly, the realizations of such a process are extremely irregular and typically do not describe realistic physical systems.

   A class of processes, where the dependence relation remains relatively simple and which finds widespread applications, are the so-called Markov processes. Such a process has a short memory in the sense that information about its past history is not required to predict its future if the present state of the process is known, $X(t) = x$. In other words, for a Markov process the past and the future are (conditionally) independent if the present is known. Mathematically, the process is Markovian if the conditional (transition) pdf possesses the property

$$P(x_n, t_n \mid x_{n-1}, t_{n-1}; \ldots; x_1, t_1) = P(x_n, t_n \mid x_{n-1}, t_{n-1}). \qquad (3.16)$$

Therefore, a Markov process is completely determined by the "initial" pdf $P(x_1, t_1)$ and the transition pdf $P(x_j, t_j \mid x_{j-1}, t_{j-1})$. Repeated use of Eqs. (3.16) and (3.5) leads to

$$P(x_n, t_n; x_{n-1}, t_{n-1}; \ldots; x_2, t_2; x_1, t_1) = P(x_n, t_n \mid x_{n-1}, t_{n-1}) \cdot$$
$$P(x_{n-1}, t_{n-1} \mid x_{n-2}, t_{n-2}) \cdots P(x_2, t_2 \mid x_1, t_1) \cdot P(x_1, t_1). \qquad (3.17)$$

3.2.1   *Chapman–Kolmogorov equation*

Consider the three-point joint pdf

$$P(x_3, t_3; x_2, t_2; x_1, t_1) = P(x_3, t_3 \mid x_2, t_2) P(x_2, t_2 \mid x_1, t_1) P(x_1, t_1). \qquad (3.18)$$

Integrating Eq. (3.18) with respect to x_2 with $t_1 < t_2 < t_3$ we find

$$\int_{\Omega_2} P(x_3, t_3; x_2, t_2; x_1, t_1) dx_2$$
$$= \int_{\Omega_2} P(x_3, t_3 \mid x_2, t_2) P(x_2, t_2 \mid x_1, t_1) P(x_1, t_1) dx_2, \qquad (3.19)$$

$$P(x_3, t_3; x_1, t_1) = \int_{\Omega_2} P(x_3, t_3 \mid x_2, t_2) P(x_2, t_2 \mid x_1, t_1) P(x_1, t_1) dx_2. \qquad (3.20)$$

Since

$$P(x_3, t_3; x_1, t_1) = P(x_3, t_3 \mid x_1, t_1) P(x_1, t_1), \qquad (3.21)$$

we conclude that the transition pdfs of a Markov process obey the equation

$$P(x_3, t_3 \mid x_1, t_1) = \int_{\Omega_2} P(x_3, t_3 \mid x_2, t_2) P(x_2, t_2 \mid x_1, t_1) dx_2, \qquad (3.22)$$

which is commonly known as the Chapman–Kolmogorov equation.[2] This equation states that for a Markov process the transition from the state $X(t_1) = x_1$ to the state $X(t_3) = x_3$ can be decomposed into two stages. First, a transition from the state x_1 at time t_1 to an arbitrary state x_2 at an arbitrary intermediate time t_2. Then, collect the transitions from all possible intermediate states to the state x_3 at time t_3.

3.2.2   *Solutions of the Chapman–Kolmogorov equation*

Equation (3.22) is a consistency equation for the conditional or transition pdfs of a Markov process and its solution gives a complete description of any Markov process. The Chapman–Kolmogorov equation is a nonlinear functional equation, and a general solution is not known. For some classes of Markov processes, either an explicit solution can be obtained, or Eq. (3.22) can be cast into the form of a partial differential equation for the transition pdf, which is easier to deal with.

We begin by considering Markov processes that are homogeneous in space and time, i.e.,

$$P(x, t \mid x', t') = P(x - x', t - t'). \qquad (3.23)$$

In this case, Eq. (3.22) can be rewritten in a form suitable for the Fourier transformation,

$$P(x - x_0, t - t_0) = \int_{-\infty}^{\infty} P(x - x', t - t') P(x' - x_0, t' - t_0) dx', \qquad (3.24)$$

where we use $\Omega_2 \equiv (-\infty, \infty)$. Recall that the Fourier transform $\tilde{P}(k, t)$ of the pdf $P(x, t)$ is know as the characteristic function. Fourier transforming Eq. (3.24), we obtain

$$\tilde{P}(k, t - t_0) = \tilde{P}(k, t - t') \tilde{P}(k, t' - t_0), \qquad (3.25)$$

---

[2]Other names for this equation are found in the literature. For example, in their book on statistical radiophysics [53], Rytov et al call Eq. (3.22) the Smoluchowski equation. In a paper by Montroll and West, this equation is called the chain equation of Bachelier, Smoluchowski, Chapman, and Kolmogorov [57]: "Equation [...] was first introduced by Bachelier (1900) in his thesis (under the direction of Poincaré) on market speculation. It was later discussed, independently by Smoluchowski (1906), Chapman (1916), and Kolmogorov (1931)." The name Smoluchowski equation is also employed in Mark Kac's famous lecture [58] and the fundamental work by Wang and Uhlenbeck [59].

which has the general solution for the conditional pdf

$$\tilde{P}(k,\tau) = \exp\left[-|k|^\beta \tau\right], \quad 0 < \beta \leq 2. \tag{3.26}$$

For $\beta = 1$, when we demand a simple symmetrical form of the solution, we obtain the Cauchy–Lorentz distribution,

$$P(x|x_0, \tau) \equiv P(x - x_0, \tau) = \frac{1}{\pi} \cdot \frac{\tau}{\tau^2 + (x - x_0)^2}, \quad \tau = D_1(t - t_0), \tag{3.27}$$

which satisfies the correct initial condition as $\tau \to 0$, namely $P(x - x_0, 0) = \delta(x - x_0)$, and D_1 is a kinetic coefficient.

If the random process is such that the second moment, or the mean squared displacement (MSD), is finite, then the solution is the Gaussian distribution with $\beta = 2$,

$$P(x|x_0, \tau) = P(x - x_0, \tau) = \frac{1}{\sqrt{4\pi\tau}} \exp\left[-\frac{(x - x_0)^2}{4\tau}\right], \quad \tau = D(t - t_0), \tag{3.28}$$

where

$$D = \lim_{t \to \infty} \frac{\langle (x - x_0)^2 \rangle}{2t} \tag{3.29}$$

is a diffusion coefficient. The MSD is

$$\langle (x - x_0)^2 \rangle = \int_{-\infty}^{\infty} (x - x_0)^2 P(x - x_0, \tau) dx$$

$$= -\frac{d^2 \tilde{P}(k, \tau)}{d k^2}\Big|_{k=0} = 2\tau = 2D(t - t_0). \tag{3.30}$$

This result corresponds to a Wiener process or Brownian motion.

For arbitrary $\beta \in (0, 2)$, the pdf $P(x - x_0, \tau)$ is expressed in the form of the inverse Fourier transformation,

$$P(x - x_0, \tau) = \frac{1}{2\pi} \int_{-\infty}^{\infty} e^{-\tau |k|^\beta} e^{-ik(x - x_0)} dk, \quad \tau = D_\beta(t - t_0). \tag{3.31}$$

It reduces to the solutions found above for $\beta = 1$ and $\beta = 2$, with the transport coefficients $D_\beta = D_1$ and $D_\beta = D$, respectively. This class of solutions is known as Lévy or stable distributions.[3] The asymptotic behavior of this expression has been studied in great detail by Montroll

---

[3]This class of solutions was investigated by Paul Lévy [60]. A random variable X with pdf P is (strictly) stable if the linear superposition $cX = aX_1 + bX_2$, where X_1 and X_2 are two independent copies of X and a, b, and c are positive constants, has the same pdf P as X. Alternatively, this property can be expressed in terms of the characteristic functions of the distribution: $\tilde{P}(ck) = \tilde{P}(ak_1)\tilde{P}(bk_2)$. A detailed discussion of stable distributions can be found, for example, in Ref. [61]; see also [3].

and West [57]. Here we present an exact expression in the form of the Fox H-function.

Example 3.1. We represent the exponential function in the form of the Mellin transform for the Fox function, as in Example 2.7 according Eq. (2.43),

$$e^{-\tau|k|^\beta} = \frac{1}{2\pi i} \int_{c-i\infty}^{c+i\infty} \Gamma(s) \left[\tau|k|^\beta\right]^{-s} ds. \tag{3.32}$$

We then perform the inverse Fourier transformation with the condition $0 < \Re(c\beta) < 1$, and find using a standard integral table [62],

$$\frac{1}{\pi} \int_0^\infty k^{-s\beta} \cos[k(x-x_0)]dk = \frac{\pi|x-x_0|^{s\beta-1}}{2\Gamma(\beta s)\cos(s\beta\pi/2)}. \tag{3.33}$$

The same result is obtained for $x - x_0 \gtrless 0$, and we take the absolute value of the argument $|x - x_0|$. Using Eq. (2.27) with shifting the index by $1/2$, we have

$$\frac{\pi}{\cos(s\beta\pi/2)} = \Gamma\left(\frac{1+s\beta}{2}\right)\Gamma\left(\frac{1-s\beta}{2}\right). \tag{3.34}$$

Substituting this result into the Mellin integration in Eq. (3.32) and using the Legendre duplication formula (2.21), $\Gamma(p)\Gamma(p+1/2) = 2^{1-2p}\sqrt{\pi}\,\Gamma(2p)$, we obtain the Mellin representation of the Fox function

$$
P(x-x_0,\tau) = \frac{|x-x_0|^{-1}}{2\pi i} \int_{c-i\infty}^{c+i\infty} ds \, \frac{\Gamma(s)\Gamma\left(\frac{1-s\beta}{2}\right)}{\Gamma\left(\frac{s\beta}{2}\right)\Gamma(1/2)} \left(\frac{2^\beta\tau}{|x-x_0|^\beta}\right)^{-s}
$$

$$
= \frac{|x-x_0|^{-1}}{\sqrt{\pi}} H_{2,1}^{1,1}\left[\tau\left(\frac{2}{|x-x_0|}\right)^\beta \left|\begin{array}{c}(\frac{1}{2},\frac{\beta}{2}),(0,\frac{\beta}{2})\\(0,1)\end{array}\right.\right]. \tag{3.35}
$$

In essence, we have inferred the Mellin-cosine transform of the Fox H-function in Eq. (2.54). Taking into account another identity of the Fox function, which also follows from its definition (2.39) [44],

$$
H_{p,q}^{m,n}\left[z \left|\begin{array}{c}(a_p,A_p)\\(b_q,B_q)\end{array}\right.\right] = H_{q,p}^{n,m}\left[\frac{1}{z}\left|\begin{array}{c}(1-b_q,B_q)\\(1-a_p,A_p)\end{array}\right.\right], \tag{3.36}
$$

we obtain the solution (3.35) in the following form,

$$
P(x-x_0,\tau) = \frac{|x-x_0|^{-1}}{\sqrt{\pi}} H_{1,2}^{1,1}\left[\tau^{-1}\left(\frac{|x-x_0|}{2}\right)^\beta \left|\begin{array}{c}(1,1)\\(\frac{1}{2},\frac{\beta}{2}),(1,\frac{\beta}{2})\end{array}\right.\right]. \tag{3.37}
$$

$\square$

3.2.3   Fokker–Planck equation

We now consider the second possibility mentioned at the beginning of Sec. 3.2.2, namely casting the Chapman–Kolmogorov equation (3.22) in the form of a partial differential equation. This can be done for an important class of Markov processes, the so-called Markovian diffusion processes. These processes look locally in time like Brownian motion or a Wiener process with a systematic drift component. Markovian diffusion processes have (almost surely) continuous sample paths or realizations. For a more detailed discussion of diffusion processes, see, e.g., Refs. [63, 64].

A Markov process $X(t)$ belongs to the class of diffusion processes, if it fulfills the following three conditions:

(1) For every positive ϵ

$$\lim_{t \to s} \frac{1}{t-s} \int_{|y-x|>\epsilon} P(y,t \mid x,s) dy = 0. \qquad (3.38)$$

This condition ensures that the Markov process has (almost surely) continuous realizations.

(2) For every positive ϵ there exist a function $A(x,s)$ such that

$$\lim_{t \to s} \frac{1}{t-s} \int_{|y-x|\leq\epsilon} (y-x) P(y,t \mid x,s) dy = A(x,s), \qquad (3.39)$$

which is an average velocity of the state change at time s and position x, or a systematic drift coefficient.

(3) For every positive ϵ there exist a function $B(x,s)$ such that

$$\lim_{t \to s} \frac{1}{t-s} \int_{|y-x|\leq\epsilon} (y-x)^2 P(y,t \mid x,s) dy = B(x,s), \qquad (3.40)$$

which is called the diffusion coefficient and describes a local variance around the systematic motion conditioned on $X(s) = x$.

If the Markovian diffusion process is homogeneous in time, i.e., $P(y,t \mid x,s) = P(y,t-s \mid x)$, then the drift and diffusion are time-independent functions.

The Chapman–Kolmogorov equation (3.22) can be cast in to the form of a partial differential equation for the transition pdf for Markovian diffusion processes:

$$\frac{\partial P(x,t|x_0,t_0)}{\partial t} + \frac{\partial A(x,t)P(x,t|x_0,t_0)}{\partial x} - \frac{1}{2}\frac{\partial^2 B(x,t)P(x,t|x_0,t_0)}{\partial x^2} = 0, \qquad (3.41)$$

which is known as the Fokker–Planck equation. The derivation proceeds as follows. We multiply the Chapman–Kolmogorov equation (3.22) by an arbitrary function $R(x)$, which vanishes identically outside a certain bounded interval:

$$\int_{-\infty}^{\infty} R(x)P(x,t|x_0,t_0)dx$$
$$= \int_{-\infty}^{\infty} P(y,t-\Delta t|x_0,t_0)dy \int_{-\infty}^{\infty} R(x)P(x|y,\Delta t)dx. \quad (3.42)$$

On the right hand side, we carry out a Taylor expansion of the function $R(x)$ with respect to y, $R(x) = R(y) + R'(y)(x-y) + \frac{1}{2}R''(y)(x-y)^2 + O\left((x-y)^3\right)$. For the first term of the expansion $R(y)$, the integration with respect to x yields 1. We move this term to the left-hand side of the equation and replace the integration variable y by x and then divide the equation by Δt. In the limit $\Delta t \to 0$, Eqs. (3.38), (3.39), and (3.40) yield

$$\int_{-\infty}^{\infty} R(x)\frac{\partial P(x,t|x_0,t_0)}{\partial t}dx$$
$$= \int_{-\infty}^{\infty} P(y,t|x_0,t_0)\left[A(y,t)R'(y) + \frac{B(y,t)}{2}R''(y)\right]dy \quad (3.43)$$

(Conditions (3.38) – (3.40) imply that the term with $O\left((x-y)^3\right)$ vanishes [65,63,64].) Integrating by parts on the right hand side of the equation and replacing y by x, we obtain

$$\int_{-\infty}^{\infty} \left[\frac{\partial P(x,t|x_0,t_0)}{\partial t} + \frac{\partial A(x,t)P(x,t|x_0,t_0)}{\partial x}\right.$$
$$\left. - \frac{1}{2}\frac{\partial^2 B(x,t)P(x,t|x_0,t_0)}{\partial x^2}\right]R(x)dx = 0. \quad (3.44)$$

Taking into account that $R(x)$ is arbitrary, we obtain a parabolic differential equation for the transition pdf $P = P(x,t|x_0,t_0)$, namely the Fokker–Planck equation (3.41).

3.3 Stochastic differential equations

Brownian motion is often modeled by the Wiener process discussed in Sec. 3.2.2. The Wiener process $W(t)$ is a zero-mean Gaussian process with stationary independent increments,

$$\langle W(t)\rangle = 0, \quad \langle W(t)W(t')\rangle = \min(t,t'). \quad (3.45)$$

The realizations of the Wiener process are (almost surely) continuous, but they are extremely "irregular." They are nowhere differentiable, i.e., the velocity of a Brownian particle is undefined, and they have infinite length on any finite time interval. An alternative model, with a well-defined velocity for the Brownian particle, is due to Langevin [66, 67],

$$m\dot{V}(t) = -\gamma V(t) + \sigma\xi(t), \tag{3.46}$$

or

$$\dot{V}(t) = -\frac{\gamma}{m}V(t) + \frac{\sigma}{m}\xi(t). \tag{3.47}$$

Here $V(t)$ is the velocity of the Brownian particle, m is the mass of the particle, and γ is the damping or friction constant. The zero-mean random force $\xi(t)$ describes the random impact of the collisions between the fluid molecules and the Brownian particle, and σ is the strength of this force. (The systematic or average impact of the collisions is accounted for by the friction force.) The random force varies on a much faster time scale than that given by m/γ, and its correlation function decreases rapidly. It is therefore common to consider the limiting case of a memoryless force, i.e., to assume that $\xi(t)$ is white noise, which has the correlation function

$$\langle\xi(t)\xi(t')\rangle = C(t - t') = \delta(t - t'), \tag{3.48}$$

where $\delta(u)$ is the Dirac delta-function. The correlation time of white noise is zero. The frequency spectrum of white noise, given by

$$S(f) = \frac{1}{2\pi}\int_{-\infty}^{\infty}\exp(-ift)C(t)dt = \frac{1}{2\pi}, \tag{3.49}$$

is completely flat. Rapidly varying real noise has a broad-band spectrum that is nearly flat. The white noise idealization replaces this spectrum by a completely flat one. In view of the Central Limit Theorem, it is further common to assume that the white noise is Gaussian distributed. Gaussian white noise (GWN) is a generalized random process. GWN can be understood as the derivative, in the generalized functions sense, of the Wiener process $W(t)$, $W(t) = \int_0^t \xi(t')dt'$. Consequently, we can write the Langevin equation (3.47) as a stochastic differential equation (SDE),

$$dV(t) = -\frac{\gamma}{m}V(t)dt + \frac{\sigma}{m}dW(t). \tag{3.50}$$

It turns out that there exists a close connection between general Langevin or stochastic differential equations driven by GWN and Markovian diffusion processes. The general one-variable Langevin equation reads

$$\dot{X}(t) = f[X(t)] + \sigma g[X(t)]\xi(t), \tag{3.51}$$

which should be understood as the SDE,

$$dX(t) = f[X(t)]dt + \sigma g[X(t)]dW(t), \tag{3.52}$$

which in turn should be understood as short-hand for the integral form,

$$X(t) = X_0 + \int_0^t f[X(t')]dt' + \sigma \int_0^t g[X(t')]dW(t'). \tag{3.53}$$

As mentioned above, the sample paths of the Wiener process $W(t)$ are very irregular. They are nowhere differentiable and have an infinite length on any finite time interval. If Eq. (3.52) has any meaning, the solution $X(t)$ will inherit the features of $W(t)$. If the noise term $g(x)$ is a constant, which may be chosen to be one, i.e., we are dealing with additive noise, then the second integral in Eq. (3.52) is trivially $\sigma W(t)$. However, if $g(x)$ does depend explicitly on the random process $X(t)$, i.e., we are dealing with multiplicative noise, then the second integral has to be dealt with carefully. It cannot be interpreted as an ordinary Riemann integral. Various formalisms have been developed to deal with stochastic integrals of the Wiener process, the most widely used being the Ito and the Stratonovich calculus [68, 69, 63, 55].

It is not our purpose here to discuss the mathematical details and subtleties of stochastic calculus.[4] A simple way of understanding the various interpretations of SDEs with multiplicative noise is to consider the implicit difference scheme [71]

$$\delta X = X(t + \tau) - X(t) = f[X(t)]\tau + \sigma g[X(t + \eta\tau)]\delta W, \tag{3.54}$$

where τ is a small time, $\eta \in [0, 1]$, and $\delta W = W(t+\tau) - W(t)$. We expand $g[X(t + \eta\tau))$ up to first order in τ and obtain with linear accuracy in τ,

$$\delta X = f[X(t)]\tau + \sigma g[X(t)]\delta W + \eta\sigma^2 g'[X(t)]g[X(t)](\delta W)^2, \tag{3.55}$$

where the prime denotes differentiation with respect to the argument. This difference scheme, where the noise term is interpreted by means of the parameter η, completely specifies the meaning of (3.52) [71]. The choices $\eta = 0, \frac{1}{2}, 1$ correspond the Ito [68], Stratonovich [69], and Hänggi–Klimontovich [72–74] interpretation of Eq. (3.52), respectively. The Stratonovich integral obeys the usual rules of calculus, whereas the Ito integral does not. The following example illustrates this point.

---

[4]For a more detailed discussion of stochastic differential equations and diffusion processes, see, e.g., Refs. [63, 64, 70].

Example 3.2. We consider the integral

$$\int_0^t W(t')dW(t').$$ (3.56)

Using the Stratonovich interpretation we obtain

$$\int_0^t W(t')dW(t') = \frac{1}{2}W(t)^2,$$ (3.57)

whereas the Ito interpretation yields

$$\int_0^t W(t')dW(t') = \frac{1}{2}\left[W(t)^2 - t\right].$$ (3.58)

$\square$

The parameter η is part of the model and must be chosen on physical grounds. There are two guidelines available. If Eq. (3.52) is obtained as the continuous-time limit of a discrete-time description, the SDE should be interpreted as an Ito SDE. On the other hand, if the noise in the SDE represents an approximation of a "real" noise, the Wong–Zakai theorem [75] implies that the SDE should be interpreted as a Stratonovich equation.

It turns out that for any interpretation $\eta \in (0,1]$ of the SDE (3.52), there exists an equivalent Ito SDE

$$(\text{Ito}) \ dX(t) = \bar{f}[X(t)]dt + \sigma g[X(t)]dW(t),$$ (3.59)

with

$$\bar{f}[X(t)] = f[X(t)] + \eta\sigma^2 g'[X(t)]g[X(t)],$$ (3.60)

in the sense that the solutions of Eqs. (3.52) and (3.59) have the same probabilistic properties. This is an important result, since the Ito stochastic integral has many desirable mathematical properties, e.g., the martingale property, not shared by any of the other possible choices. The Ito calculus is the most well developed and complete stochastic calculus, both from an analytical as well as a numerical viewpoint. To quote Gardiner [55, p. 92]: "The Ito integral is mathematically and technically the most satisfactory, but unfortunately, it is not always the most natural choice physically."

The solutions of Ito SDEs are Markovian diffusion processes. The solutions of general SDEs are Markovian if and only if the noise is white, i.e., it is the time derivative, in the generalized functions sense, of a process with stationary independent increments on non-overlapping intervals. The class of processes with stationary independent increments, and therefore the class of white noises, is completely known. Extended Fokker–Planck equations for SDEs with non-Gaussian white noise have been derived, see, e.g., Ref. [76]. If the white noise is Gaussian, then the solution of the SDE is a Markovian diffusion process. The pdf of the solution $X(t)$ of the Ito SDE (3.59) obeys the Fokker–Planck equation

$$\frac{\partial P(x,t)}{\partial t} = -\frac{\partial}{\partial x}\bar{f}(x)P(x,t) + \frac{\sigma^2}{2}\frac{\partial^2}{\partial x^2}g^2(x)P(x,t), \qquad (3.61)$$

i.e., $A(x,t) = \bar{f}(x)$ and $B(x,t) = \sigma^2 g^2(x)$. The stationary solution of Eq. (3.61), if it exists, is given by

$$P_s(x) = \frac{N}{g^2(x)}\exp\left[\frac{2}{\sigma^2}\int^x \frac{\bar{f}(y)}{g^2(y)}dy\right], \qquad (3.62)$$

where N is the normalization constant.

3.4 Markov chains

The Chapman–Kolmogorov equation can be written also for a process $X(t)$ with discrete state space and discrete time t. In particular, the Eqs. (3.6) and (3.22) for the pdf $P(x_n,t) \equiv P(n,t)$ and the transition pdf $P(x_n|x_m,t) \equiv P(n|m,t)$ read now

$$P(n,t) = \sum_m P(n,t|m,t-1)P(m,t-1), \qquad (3.63a)$$

$$P(n,t|n_0,t_0) = \sum_m P(n,t|m,t-1)P(m,t-1|n_0,t_0), \qquad (3.63b)$$

respectively. If the state space is finite or countable, the process is called a Markov chain, and Eq. (3.63) is known as the Markov equation. The transition probability $P(n,t|m,t-1)$ contains the necessary information about the transition mechanism from the state m to n. One can thus define the *transition matrix* $\mathbf{Q}(t-1)$ whose (m,n)-th element is

$$Q_{m,n} = P(n,t|m,t-1). \qquad (3.64)$$

For a time-homogeneous process, the matrix is time independent, and the probability that $X(t)$ has a value x_n at time t is

$$P(n,t) = \sum_m Q_{n,m}P(m,t-1). \qquad (3.65)$$

A more detailed discussion of Markov chains and their applications can be found, e.g., in the book by Karlin and Taylor [54]. Here we consider only some examples of random walks on the one dimensional lattice.

3.4.1   *The simple random walk*

Let us consider Eq. (3.65) for a random walk in one dimension, where a particle moves to the right with probability p and to the left with probability $q = 1 - p$. Then the elements of the transition matrix are

$$Q_{m,n} = p\delta_{n,m+1} + q\delta_{n,m-1},\qquad(3.66)$$

and Eq. (3.63b) reads

$$P(n,t) = pP(n-1,t-1) + qP(n+1,t-1),\qquad(3.67)$$

where $t = 1, 2, \ldots$ is the number of steps taken by the random walker. The particle starts at the site l, i.e., the initial condition is $P(n,0) = \delta_{n,l}$.

We apply again the method of characteristic functions, which is the discrete Fourier transform on the interval $[-\pi, \pi]$,

$$\tilde{P}(\theta,t) = \sum_{n=-\infty}^{\infty} e^{in\theta}P(t,n), \quad P(n,t) = \frac{1}{2\pi}\int_{-\pi}^{\pi} e^{-in\theta}\tilde{P}(\theta,t)d\theta.$$

$$(3.68)$$

Equations (3.66) and (3.68) imply that the characteristic function is given by

$$\tilde{P}(\theta,t) = (pe^{i\theta} + qe^{-i\theta})^t \tilde{P}(\theta,0),\qquad(3.69)$$

where $\tilde{P}(\theta,0) = e^{il\theta}$. Consequently the solution for the pdf reads

$$P(n,t) = \frac{1}{2\pi}\int_{-\pi}^{\pi}\sum_{j=1}^{t}\binom{t}{j}p^j q^{t-j}e^{i\theta(2j-t-n+l)}d\theta = \binom{t}{\frac{t+n-l}{2}}p^{\frac{t+n-l}{2}}q^{\frac{t-n+l}{2}},$$

$$(3.70)$$

where $\binom{n}{m} = \frac{n!}{m!(n-m)!}$ is the binomial coefficient.

Example 3.3. In the asymptotic limit $t \to \infty$, the binomial distribution (3.70) is approximately a Gaussian distribution, which can be immediately obtained from Stirling's formula for the gamma function $\Gamma(z)$,

$$\Gamma(z) \approx \sqrt{\frac{2\pi}{z}} e^{-z} z^z. \tag{3.71}$$

Therefore for $p = q = 1/2$, the solution (3.70) reduces to the Gaussian distribution

$$P(n,t) \approx \frac{1}{\sqrt{\pi t}} e^{\frac{(n-l)^2}{2t}}. \tag{3.72}$$

In the case of general p and q, this asymptotic behavior is known as the de Moivre–Laplace theorem [53], which is a special case of the Central Limit Theorem. □

3.4.2 The first passage problem

An important characteristic of a random walk is the first-passage probability,[5] which is the probability that a particle visits a point n for the first time at step t. For our example of the discrete random walk in one dimension, the first-passage probability $F(n,t)$ is defined in complete analogy with the probability $P(n,t)$. These two quantities are related to each other as follows

$$P(n,t) = \delta_{t,0}\delta_{n,0} + \sum_{t'=1}^{t} F(n,t')P(0,t-t'). \tag{3.73}$$

The rhs of the expression has the following interpretation: the first term determines the initial condition of a particle which starts at the origin $n = 0$ at step 0. After t steps it finishes at the point n. However, it can visit the point n at some earlier step t' with the probability $F(n,t')$ and for the other $t - t'$ steps it makes a loop (or loops) by leaving and returning to n with the transition probability $P(n - n, t - t') = P(0, t - t')$.

To find $F(n,t)$, we use a *generating function*, or z-transform, which can be introduced by analogy with the characteristic function. We multiply Eq. (3.73) by z^t and sum over t,

$$\sum_{t} P(n,t)z^t = \delta_{n,0} + \sum_{t} z^t \sum_{t'=1}^{t} F(n,t')P(0,t-t'). \tag{3.74}$$

---

[5]A complete description can be found in Refs. [77, 37]. We follow the details of these pedagogically excellent books.

Using the definitions of the z-transforms,

$$P(n,z) = \sum_t P(n,t)z^t, \qquad F(n,z) = \sum_t F(n,t)z^t, \qquad (3.75)$$

and using the convolution rule, also known as the Cauchy rule,

$$\sum_{k=0}^{\infty}\sum_{l=0}^{k} a_l b_{k-l} = \left(\sum_{k=0}^{\infty} a_k\right)\left(\sum_{k=0}^{\infty} b_k\right), \qquad (3.76)$$

we arrive at the expression

$$F(n,z) = \frac{P(n,z) - \delta_{n,0}}{P(0,z)}. \qquad (3.77)$$

We have reduced the problem to finding the generating function $P(n,z)$. To that end, we return to the simplified Eq. (3.66) with $p = q = 1/2$ and the initial condition $P(n,0) = \delta_{n,0}$. Performing both the Fourier and z-transforms, we obtain from Eq. (3.66)

$$\tilde{P}(\theta,z) = z\cos(\theta)\tilde{P}(\theta,z) + 1, \qquad (3.78)$$

with the solution in the form of the integral

$$P(n,z) = \frac{1}{2\pi}\int_{-\pi}^{\pi} \frac{e^{-i\theta n}\,d\theta}{1 - z\cos(\theta)}. \qquad (3.79)$$

The result of the integration [77] is discussed in the next example.

Example 3.4. We perform the variable change $w = e^{\mp i\theta}$ for $n \geqslant 0$. Due to symmetry, the result is the same for both positive and negative n, and we take $e^{-i\theta n} = w^{|n|}$. In the complex w-plane the integral (3.79) corresponds to the contour integral with a contour around a unit circle:

$$P(n,z) = \frac{1}{2\pi i z}\oint \frac{w^{|n|}\,dw}{(w - w_1)(w - w_2)}, \qquad (3.80)$$

where $w_{1,2} = \frac{1}{z}(1 \pm \sqrt{1 - z^2})$ are the roots of the quadratic equation. Since $|w_2| < 1$, it is the only pole of the integrand, and the result of the residue calculation is

$$P(n,z) = \frac{\left[\frac{1}{z}\left(1 - \sqrt{1 - z^2}\right)\right]^{|n|}}{\sqrt{1 - z^2}}. \qquad (3.81)$$

$\square$

The probability of the walker reaching n at some time during the random walk is

$$F(n) = \sum_{t=0}^{\infty} F(n,t) = F(n, z = 1).$$ (3.82)

Consequently, the probability for the walker to return to the origin is given by

$$F(0) = \sum_{t=0}^{\infty} F(0,t) = F(0, z = 1).$$ (3.83)

According to Eq. (3.77)

$$F(0,z) = 1 - \frac{1}{P(0,z)}.$$ (3.84)

Let us define $\chi(t) = 1$ if $X(t) = 0$ and $\chi(t) = 0$ otherwise. Then $N = \sum_t \chi(t)$ is the number of times that the walker is at the origin, $n = 0$. The average value of N is given by

$$\langle N \rangle = \left\langle \sum_t \chi(t) \right\rangle = \sum_t \langle \chi(t) \rangle = \sum_t \sum_n \chi(t) P(n,t)$$

$$= \sum_t P(0,t) = P(0, z = 1).$$ (3.85)

Equation (3.72) implies that $P(0,t) \approx \frac{1}{\sqrt{\pi t}}$ and consequently that $P(0, z = 1) = \sum_t P(0,t) = \infty$. In other words, the walker visits the origin on average infinitely often. Substituting this result into Eq. (3.84), we find that $F(0, z = 1) = 1$. The nearest-neighbor lattice random walk in one dimension is recurrent, since the probability to return to the origin is one. In general, any one dimensional random walk is recurrent.[6]

The question then arises as to the distribution of number of steps for this recurrent walk, or what is the *first return probability* to the origin $F(0,t)$. This question is answered in the next example.

Example 3.5. Using Eqs. (3.75) and (3.81), we find

$$\sum_{t=0}^{\infty} F(0,t) z^t = 1 - \sqrt{1 - z^2}.$$ (3.86)

We write the rhs as the binomial series,

$$1 - \sqrt{1 - z^2} = -\sum_{k=1}^{\infty} \frac{\Gamma\left(\frac{3}{2}\right)(-1)^k}{\Gamma(k+1)\Gamma\left(\frac{3}{2} - k\right)} z^{2k}.$$ (3.87)

---

[6]See the discussion in Ref. [37].

Taking into account the relation for the Gamma function,

$$\frac{-1}{\Gamma\left(\frac{3}{2}-k\right)} = \frac{2\cos(\pi k)\Gamma\left(\frac{1}{2}+k\right)}{\pi(2k-1)}, \tag{3.88}$$

and using the Stirling's formula, see Example 3.3, we obtain

$$F(0,t) = \begin{cases} \frac{1}{2}\sqrt{\frac{e}{\pi}}k^{-\frac{3}{2}} & t \text{ is even, } t = 2k \\ 0 & t \text{ is odd.} \end{cases} \tag{3.89}$$

The last line is obvious, since the return to the origin in one dimension is possible only for an even number of steps. Of course, the first passage through the origin coincides with the first return to the origin for the nearest-neighbor lattice random walk in one dimension.   □

The result in Eq. (3.89) is a special case of the Sparre Andersen theorem [78]. The latter is discussed in Ref. [37], and we quote: "For any symmetric random walk (independent of the existence of moments) the first passage probability through the origin behaves [. . .] asymptotically" as

$$F(0,t) = \frac{1}{2\sqrt{\pi}}t^{-3/2}. \tag{3.90}$$

3.5   Statistics of recurrence times as a Markov process on a comb

The one-dimensional Pólya problem [79] on the recurrence of a Markov random walk to the origin, discussed in the previous section, leads to another interesting question on the statistics of the recurrence times. Considering this question, we follow a result by Havlin and Weiss [13, 35] to obtain an explicit expression for the waiting time probability $\psi(t)$ for a comb model depicted in Fig. 1.1. The random walk is described in terms of the first passage probability $F(n,t)$ to be at site n for the first time after t steps (or at discrete time t), or its generating function $F(n,z)$.

We consider the case of a comb, where the walker moves in the two-dimensional (x, y)-space and where the y-axis represents the Markov process in the fingers with the origin at the backbone $(x, 0)$. To describe this walk, we introduce the probability $F^{(m)}(0,t)$ to be at the origin for the m-th time at step t. Therefore, the probability to be at the origin for the $(m+1)$-th time is given by the Markov chain equation

$$F^{(m+1)}(0,t) = \sum_{l=0}^{t} F^{(m)}(0,l)F(0,t-l), \tag{3.91}$$

which accounts for all possibilities to arrive at $n = 0$ again after the m-th visit at an earlier time. Performing the z-transform of Eq. (3.91), we obtain the recurrence relation for the generating functions,

$$F^{(m)}(0, z) = F^{(m-1)}(0, z)F(0, z) = [F(0, z)]^m . \qquad (3.92)$$

Taking into account expression (3.77) for $n = 0$, we have

$$F^{(m)}(0, z) = \left[1 - \frac{1}{P(0, z)}\right]^m . \qquad (3.93)$$

Let us consider the comb geometry shown in Fig. 3.1. It describes a discrete random walk in the upper half part, $y \geq 0$, of the comb.[7] The random walker along the fingers jumps to the nearest site with probability $1/2$. Reaching the coordinates $(x, 0)$, the walker can either move along the backbone, i.e., along the x-axis in the positive or negative direction, with probability $1/2$, or return to the finger with probability $1/2$, either in the positive or negative y-direction. Therefore a random walk along the backbone $(x, 0)$ is interrupted by walks inside the fingers. The durations of these interruptions are random times or number of steps in our discrete dynamics. The delay time probability $\psi(t)$ is the probability[8] that the waiting time between any two successive jumps along the backbone is equal to t steps. To obtain the waiting time probability, or the pdf in the continuous limit, we need to account for all possible visits to the backbone at the junction $(x, 0)$ according to Eqs. (3.73) and (3.75). Note that we assume that the length of the fingers goes to infinity, $l \to \infty$.

Let us consider the m-th visit of the origin of the fingers at point A, see Fig. 3.1, which is a jump from point A' at step t. One possibility, which occurs with probability $1/2$, is that the walker jumps from A' to A. Such a realization contributes $\frac{1}{2}\delta_{t,1}$ to $\psi(t)$. Another possibility is that the random walker jumps back to point A'' with the probability $1/2$, and then returns to the backbone at random step t for the first time. This realization contributes $\frac{1}{4}F^{(1)}(0, t)$ to $\psi(t)$. The walker can now move along the backbone, in either direction, with probability $1/2$, or move into the fingers with the probability $1/2$. In the second case, the walker returns to the backbone for the second time at the time t with the probability $\frac{1}{8}F^{(2)}(0, t)$. This process can be repeated *ad infinitum*. Taking into account

---

[7]We suppose that the comb dynamics in the fingers is symmetric with respect to the backbone.

[8]In the continuous time random walk (CTRW), $\psi(t)$ is the waiting time pdf.

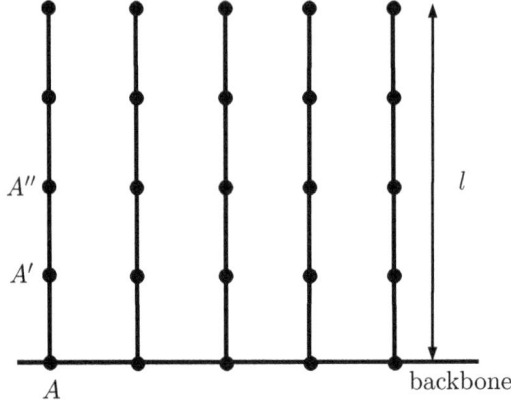

Fig. 3.1   Discrete random walk on comb.

all these contributions to the waiting time probability, we find

$$\psi(t) = \frac{1}{2}\left[\delta_{t,1} + \sum_{m=1}^{\infty}\left(\frac{1}{2}\right)^m F^{(m)}(0,t)\right].\tag{3.94}$$

We switch to the generating function for further analysis. Performing the z-transform, we obtain from Eq. (3.94)

$$\psi(z) = \frac{z}{2} + \frac{1}{2}\sum_{m=1}^{\infty}\left(\frac{1}{2}\right)^m\left[1 - \frac{1}{P(0,z)}\right]^m = \frac{z}{2} + \frac{1}{2}\frac{P(0,z)-1}{P(0,z)+1},\tag{3.95}$$

where $P(0,z) = 1/\sqrt{1-z^2}$ is obtained in Eq. (3.81). It is convenient to introduce the survival probability, which is the probability to stay inside the fingers for at least t steps,

$$\Psi(t) = 1 - \sum_{l=0}^{t}\psi(l).\tag{3.96}$$

Its generating function reads for $|z| < 1$

$$\Psi(z) = \sum_{t=0}^{\infty}\Psi(t)z^t = \sum_{t=0}^{\infty}z^t - \sum_{t=0}^{\infty}z^t\sum_{l=0}^{t}\psi(l)$$

$$= \sum_{t=0}^{\infty}z^t\left[1 - \sum_{l=t}^{0}z^{t-l}\psi(t-l)\right] = \frac{1-\psi(z)}{1-z}.\tag{3.97}$$

Here we used $\sum_{l=0}^{t}\cdots = \sum_{l=t}^{0}\cdots$ and the Cauchy rule (3.76). The z-transform $P(0,z)$ tends to infinity in the recurrent limit as $z \to 1^-$ as follows,

$$P(0,z) = \frac{1}{\sqrt{(1-z)(1+z)}} \xrightarrow{z\to 1} \frac{1}{\sqrt{2(1-z)}}.\tag{3.98}$$

Substituting Eqs. (3.95) and (3.98) into Eq. (3.97), we obtain the survival generating function

$$\Psi(z) = \frac{1}{\sqrt{2(1-z)}} - \frac{1}{2}\sqrt{z-1}, \tag{3.99}$$

which in the recurrent limit reads

$$\Psi(z) = \sum_{t=0}^{\infty} \Psi(t)z^t \sim \frac{1}{\sqrt{2(1-z)}}. \tag{3.100}$$

This asymptotic expression is suitable for the inverse z-transform, which exactly corresponds to the discrete Tauberian theorem for the power series, see Ref. [51], Sec. XIII.5, Theorem 5. This theorem establishes a relation between the survival probability $\Psi(t)$ and the explicit form of its generating function (3.100), namely

$$\Psi(t) \sim \frac{1}{\sqrt{2\pi t}}, \qquad t \to \infty. \tag{3.101}$$

Moreover, this result also determines the waiting time probability $\psi(t)$. Since the waiting time probability $\psi(t)$ is a monotonic function or sequence, it also follows from the Tauberian theorem and Eq. (3.101) that the sum

$$\sum_{l=0}^{\infty} \psi(l) = \Psi(t) \sim \frac{1}{\sqrt{2\pi t}} \tag{3.102}$$

also implies the asymptotic expression for $\psi(t)$,

$$\psi(t) \sim t^{-\frac{3}{2}}, \qquad t \to \infty. \tag{3.103}$$

3.5.1 *Auxiliary: Generating function formalism*

For the convenience of the reader, we present some brief explanatory notes on the generating function formalism or z-transform method, which is useful for Markov chains, as described in this chapter.[9] By means of the generating function formalism, a probability function $f(t) = f_n$, defined at discrete times $t = n$, is transformed into its continuous counterpart, known as a generating function,

$$F(z) = \sum_{n=0}^{\infty} f_n z^n. \tag{3.104}$$

Here z is a complex variable, such that $|z| < R$, where R is the radius of convergence. In our considerations, $R = 1$. The advantage of the z-transform (3.104) lies in the reduction of convolutions, which involve the

---

[9]We follow Sec. 3.2 of Ref. [80] and Sec. XIII.5 of Ref. [51].

sequence $\{f_n\}$, as for example in Markov chains, to algebraic expressions for $F(z)$. The latter can be easily solved and explicit expressions for the generating function $F(z)$ can be obtained. Then the sequence $\{f_n\}$ can be obtained by the inverse z-transform,

$$f_n = \frac{1}{2\pi i} \oint_C F(z) z^{-(n+1)} dz. \tag{3.105}$$

Here C is a counterclockwise closed path encircling the origin and entirely in the region of convergence. Equation (3.105) can be obtained from Eq. (3.104) by multiplying both sides by $z^{-(n+1)}$ and integrating with respect to z. The lhs of Eq. (3.104) yields the rhs of Eq. (3.105), while for the rhs of Eq. (3.104) we obtain

$$\frac{1}{2\pi i} \sum_{m=0}^{\infty} \oint_C f_m z^{-(n-m+1)} dz = \frac{1}{2\pi} \sum_{m=0}^{\infty} f_m \frac{1}{2\pi} \int_0^{2\pi} e^{i(m-n)\theta} d\theta$$

$$= \sum_{m=0}^{\infty} f_m \delta_{m,n} = f_n. \tag{3.106}$$

Here we take the contour as a circle with the radius $R = 1$ and make the change of the variable $z = e^{i\theta}$.

In many cases, the knowledge of the explicit form of $F(z)$ does not help to calculate the integral, which can be very complicated. In that case, the discrete Tauberian theorem for power series [51] is useful, as it establishes a relation between f_n and $F(z)$.

The discrete Tauberian theorem [51]

Let $f_n \geq 0$ and suppose that the power series

$$F(z) = \sum_{n=0}^{\infty} f_n z^n \tag{3.107}$$

converges for $0 \leq z < 1$ and that

$$F(z) \sim \frac{1}{(1-z)^\rho} L\left(\frac{1}{1-z}\right), \qquad z \to 1^-, \tag{3.108}$$

where L is a slowly varying function near infinity and $0 \leq \rho < \infty$. Then Eq. (3.108) is equivalent the following two relations.

$$f_0 + f_1 + \cdots + f_{n-1} \sim \frac{1}{\Gamma(\rho+1)} n^\rho L(n), \qquad n \to \infty, \tag{3.109}$$

and

$$f_n \sim \frac{1}{\Gamma(\rho)} n^{\rho-1} L(n), \qquad n \to \infty. \tag{3.110}$$

The latter expression also supposes that $\{f_n\}$ is a monotonic sequence.

In conclusion, we note that we used both results of the theorem to infer the explicit expression for the waiting time probability $\psi(t)$. Our result in Eq. (3.101) follows from Eq. (3.110), which yields $\Psi(t)$. Comparing Eqs. (3.109) and (3.110) with Eq. (3.96) for $\Psi(t)$ and $\psi(t)$, we arrived at the expression in Eq. (3.103).

PART 2

Random Walks on Combs

Chapter 4

Mesoscopic description of random walks on combs: CTRW

4.1 Mesoscopic description

We present a simple example of random walks on combs with a one-dimensional backbone and perpendicular branches to explain the CTRW mechanism of the waiting times along the backbone. The statistical properties of comb-like structures have been widely studied. In particular, the first passage-time and the survival probability have been analyzed, for example, in Refs. [81, 77, 13]. Interest has also focused on the waiting time distribution, which is equivalent to performing a random walk along the branches [81, 29, 30, 82, 83], the mean encounter time between two random walkers [32], and the occupation time statistics [33].

To be specific, we consider the following system. The walker moves through the comb by performing jumps between nearest-neighbor nodes along the backbone or along the branches. We assume that the walker performs isotropic jumps along the backbone, but the jumps along the branches may be biased, for example by an external field [12]. We derive the balance equation for the pdf $P_1(x, t)$ of finding the walker at node x on the backbone at time t. When the walker arrives at a node, it waits a random time τ before performing a new jump to a nearest-neighbor node. We assume that the comb is homogeneous, and the local waiting time pdf at any given node is described by $\phi_0(\tau)$. When the walker enters a branch, it spends some time moving inside the branch before returning to the backbone. This sojourn time can be used to determine an effective waiting time pdf $\psi(\tau)$ for the walker's motion along the backbone. In other words, the motion of the walker on the comb can be reduced to an effective motion along a one-dimensional lattice, corresponding to the backbone only. This motion is non-Markovian and can be described mesoscopically by the Generalized

Master Equation (GME) for the pdf $P_{\text{bb}}(x,t)$ of finding the walker at node x on the backbone at time t:

$$\frac{\partial P_{\text{bb}}}{\partial t} = \int_0^t K(t-t') \left[\int_{-\infty}^{\infty} P_{\text{bb}}(x-x',t')Q(x')dx' - P_{\text{bb}}(x,t') \right] dt'. \quad (4.1)$$

Here $K(t)$ is the memory kernel related to the waiting time pdf via its Laplace transform, $\hat{K}(s) = s\hat{\psi}(s)/[1 - \hat{\psi}(s)]$, according Eq. (1.24), where s is the Laplace variable. The dispersal kernel $Q(x)$ is a generalization of the transition matrix $Q_{m,n}$ of a Markov chain, see Eq. (3.66). It represents the probability of the walker performing a jump of length x. If the walker moves isotropically between nearest neighbors in a one-dimensional lattice of spacing a, the dispersal kernel reads

$$Q(x) = \delta(x-a)/2 + \delta(x+a)/2. \quad (4.2)$$

We assume that the walker is initially located at $x = 0$, i.e., $P_{\text{bb}}(x,0) = \delta_{x,0}$, with $x = ia$ and $i = 0, \pm 1, \pm 2, \ldots$, where $\delta_{x,0}$ is the Kronecker delta. Then the Laplace transform of the GME for $x \neq 0$ reads

$$\hat{P}_{\text{bb}}(x,s) = \frac{\hat{\psi}(s)}{2} \left[\hat{P}_{\text{bb}}(x-a,s) + \hat{P}_{\text{bb}}(x+a,s) \right]. \quad (4.3)$$

To derive the effective waiting time pdf $\psi(t)$ for the backbone dynamics and relate it to the local waiting time pdf $\phi_0(t)$ and the other mesoscopic characteristics of the comb, viz. q and N, we formulate the mesoscopic balance equation for the CTRW on the comb. Let $P(x,y,t)$ be the pdf that the walker on the comb is located at the node with backbone coordinate x and branch coordinate y at time t, and let $\hat{P}(x,y,s)$ be its Laplace transform. Taking into account the contributions of the walker arriving from the upper and the lower branch (see Fig. 4.1), we obtain the mesoscopic balance equation for the walker being at node $(x,y) = (ia, 0)$, i.e., being at a backbone node,

$$\hat{P}(x,s) = \frac{\hat{\phi}_0(s)}{4} \left[\hat{P}(x-a,s) + \hat{P}(x+a,s) \right]$$
$$+ (1-q)\hat{\phi}_0(s) \left[\hat{P}(y=a,s) + \hat{P}(y=-a,s) \right]. \quad (4.4)$$

To achieve a concise notation, we are using $\hat{P}(x,s)$, $\hat{P}(x-a,s)$, and $\hat{P}(x+a,s)$ as shorthand for $\hat{P}(x,y=0,s)$, $\hat{P}(x-a,y=0,s)$, and $\hat{P}(x+a,y=0,s)$, while $\hat{P}(y=a,s)$ and $\hat{P}(y=-a,s)$ stand for $\hat{P}(x,y=a,s)$ and $\hat{P}(x,y=-a,s)$, respectively.

The term

$$\frac{\hat{\phi}_0(s)}{4} \left[\hat{P}(x-a,s) + \hat{P}(x+a,s) \right] \quad (4.5)$$

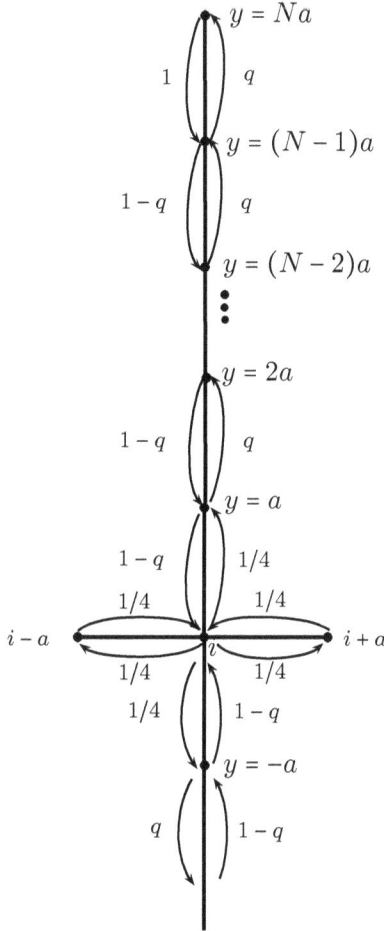

Fig. 4.1   Schematic representation of the possible jumps of a walker with the corresponding probabilities. Reprinted figure with permission from V. Méndez, A. Iomin, D. Campos, and W. Horsthemke, Phys. Rev. E, 92, 062112 (2015). Copyright 2015 by the American Physical Society.

corresponds to the contribution of the walker arriving at node $x = ia$ from the left or from the right with probability $1/4$, after waiting a random time τ with pdf $\phi_0(\tau)$ at nodes $x + a$ or $x - a$.

As shown in Fig. 4.1, the walker located at the ith node of the backbone may jump to the right, left, up, or down with probability $1/4$. We assume that the walker moves forward (away from the backbone) along the branches with probability q and back to the backbone with probability $1 - q$. The term

$$(1 - q)\hat{\phi}_0(s) \left[\hat{P}(y = a, s) + \hat{P}(y = -a, s) \right] \qquad (4.6)$$

in Eq. (4.4) corresponds the contribution of the walker arriving at the backbone node x from the first node of the upper or lower branch after waiting there a random time τ with pdf $\phi_0(\tau)$.

If we can express $\hat{P}(y = a, s)$ and $\hat{P}(y = -a, s)$ in Eq. (4.4) in terms of $\hat{P}(x, s)$, then Eq. (4.4) can be cast in the form of Eq. (4.3). In other words, any contribution from branch nodes will have been eliminated, and we will obtain a closed balance equation for $\hat{P}(x, s) = \hat{P}(x, y = 0, s)$ purely in terms of the probabilities of the walker being at adjacent backbone nodes. We can then identify $\hat{P}(x, s)$ with $\hat{P}_{bb}(x, s)$ and replace the CTRW on the comb by an effective CTRW on a one-dimensional lattice, corresponding to the backbone. This effective walk accounts for the excursions of the walker on the comb into the side branches in terms of an effective waiting time pdf $\psi(\tau)$ at backbone nodes.

We proceed as follows. Consider the motion along the upper branches. The lower branch dynamics is the same due to the symmetry of the comb. The mesoscopic balance equation for the first node of the upper branches reads

$$\hat{P}(y = a, s) = \frac{\hat{\phi}_0(s)}{4} \hat{P}(x, s) + \hat{\phi}_0(s)(1 - q)\hat{P}(y = 2a, s). \qquad (4.7)$$

The first term $\hat{\phi}_0(s)\hat{P}(x, s)/4$ corresponds to the contribution of the walker arriving from the backbone, while $\hat{\phi}_0(s)(1 - q)\hat{P}(y = 2a, s)$ is the contribution of the walker jumping from the upper node $y = 2a$ to $y = a$ with probability $1 - q$ after waiting a random time τ with pdf $\phi_0(\tau)$. Analogously, we have for the lower branches

$$\hat{P}(y = -a, s) = \frac{\hat{\phi}_0(s)}{4} \hat{P}(x, s) + \hat{\phi}_0(s)(1 - q)\hat{P}(y = -2a, s). \qquad (4.8)$$

Generalizing (4.7) to any node of the branches located between $2a \leq y \leq (N - 2)a$, we obtain the balance equation for the upper branches

$$\hat{P}(y, s) = \phi_0(s) \left[q\hat{P}(y - a, s) + (1 - q)\hat{P}(y + a, s) \right]. \qquad (4.9)$$

A similar balance equation holds for the lower branches.

4.1.1 Laplace transform of the waiting time pdf $\hat{\psi}(s)$

To determine the Laplace transform $\hat{\psi}(s)$ of the effective backbone node waiting time pdf, we need to express $\hat{P}(y = a, s)$ and $\hat{P}(y = -a, s)$ in (4.4) in terms of $\hat{P}(x,t)$, so that Eq. (4.4) can be cast in the form of Eq. (4.3). Given Eqs. (4.7) and (4.8), this goal can be achieved if $P(y = 2a, s)$ and $P(y = -2a, s)$ can be related to $\hat{P}(y = a, s)$ and $\hat{P}(y = -a, s)$. We proceed as follows. The solution of Eq. (4.9) reads

$$\hat{P}(y, s) = A_1 \lambda_+^{y/a} + A_2 \lambda_-^{y/a}, \tag{4.10}$$

where

$$\lambda_\pm = \frac{1 \pm \sqrt{1 - 4q(1-q)\phi_0^2(s)}}{2(1-q)\phi_0(s)}. \tag{4.11}$$

Solutions for the coefficients A_1 and A_2 are considered in the example, which deals with the lower "boundary dynamics" at $y = a$ and $y = 2a$.

Example 4.1. To find expressions for the quantities A_1 and A_2, whose dependence on x and s is not displayed, we apply (4.10) to the node $y = 2a$:

$$\hat{P}(y = 2a, s) = A_1 \lambda_+^2 + A_2 \lambda_-^2. \tag{4.12}$$

On the other hand, setting $y = 2a$ in (4.9), we find

$$\hat{P}(y = 2a, s) - \hat{\phi}_0(s)q\hat{P}(y = a, s) = \hat{\phi}_0(s)(1 - q)\hat{P}(y = 3a, s). \tag{4.13}$$

Setting $y = 3a$ in solution (4.10), we obtain

$$\hat{P}(y = 2a, s) - q\hat{\phi}_0(s)\hat{P}(y = a, s) = \hat{\phi}_0(s)(1 - q)\left[A_1 \lambda_+^3 + A_2 \lambda_-^3\right]. \tag{4.14}$$

Solving the system of equations (4.12) and (4.14) for the quantities A_1 and A_2, we obtain

$$A_1 = \frac{\hat{P}(y = 2a, s) - q\hat{\phi}_0(s)\hat{P}(y = a, s)}{\lambda_+^2 \left(\lambda_+ - \lambda_-\right)\hat{\phi}_0(s)(1 - q)} - \frac{\lambda_-\hat{P}(y = 2a, s)}{\lambda_+^2 \left(\lambda_+ - \lambda_-\right)}, \tag{4.15a}$$

$$A_2 = \frac{-\hat{P}(y = 2a, s) + q\hat{\phi}_0(s)\hat{P}(y = a, s)}{\lambda_-^2 \left(\lambda_+ - \lambda_-\right)\hat{\phi}_0(s)(1 - q)} + \frac{\lambda_+\hat{P}(y = 2a, s)}{\lambda_-^2 \left(\lambda_+ - \lambda_-\right)}. \tag{4.15b}$$

□

A special situation occurs at the end of the branches, where we have to impose reflecting boundary conditions, i.e.,

$$\hat{P}(y = Na, s) = q\hat{\phi}_0(s)\hat{P}(y = (N - 1)a, s). \tag{4.16}$$

The node at $y = (N-1)a$ also needs a special balance equation [see Fig. (4.1)],

$$\hat{P}(y = (N-1)a, s) = q\hat{\phi}_0(s)\hat{P}(y = (N-2)a, s) + \hat{\phi}_0(s)\hat{P}(y = Na, s). \quad (4.17)$$

Substituting $y = (N-2)a$ into Eq. (4.9) and taking into account Eq. (4.16), we can write

$$\hat{P}(y = (N-2)a, s) = h(\hat{\phi}_0(s))\hat{P}(y = (N-3)a, s), \quad (4.18)$$

where

$$h(\hat{\phi}_0(s)) = \frac{q\hat{\phi}_0(s)\left[1 - q\hat{\phi}_0^2(s)\right]}{1 + q(q-2)\hat{\phi}_0^2(s)}. \quad (4.19)$$

Substituting the solutions from Eqs. (4.10) and (4.15) into Eq. (4.18), we find

$$\hat{P}(y = 2a, s) = \Lambda(q, \hat{\phi}_0(s))\hat{P}(y = a, s), \quad (4.20)$$

where

$$\Lambda(q, \hat{\phi}_0(s)) = \frac{2q\hat{\phi}_0(s)}{1 + \dfrac{1 + \lambda_N(q, \hat{\phi}_0(s))}{1 - \lambda_N(q, \hat{\phi}_0(s))}\sqrt{1 - 4q(1-q)\hat{\phi}_0^2(s)}}, \quad (4.21a)$$

$$\lambda_N(q, \hat{\phi}_0(s)) = \left(\frac{\lambda_-}{\lambda_+}\right)^{N-5}\frac{\lambda_- - h(\hat{\phi}_0(s))}{\lambda_+ - h(\hat{\phi}_0(s))}. \quad (4.21b)$$

For the lower branch we obtain in a similar manner,

$$\hat{P}(y = -2a, s) = \Lambda(q, \hat{\phi}_0(s))\hat{P}(y = -a, s). \quad (4.22)$$

We have achieved our goal of expressing $P(y = 2a, s)$ and $P(y = -2a, s)$ in terms of $\hat{P}(y = a, s)$ and $\hat{P}(y = -a, s)$. Substituting Eqs. (4.20) and (4.22) into Eqs. (4.7) and (4.8) and using the resulting expressions in Eq. (4.4), we obtain an equation of the form (4.3) with

$$\hat{\psi}(s) = \frac{\hat{\phi}_0(s)}{2 - \dfrac{(1-q)\hat{\phi}_0^2(s)}{1 - (1-q)\hat{\phi}_0(s)\Lambda(q, \hat{\phi}_0(s))}}. \quad (4.23)$$

The inverse Laplace transformation of (4.23) yields $\psi(t)$, which incorporates the dynamics along the branches and can be understood as the effective waiting time pdf for a walker moving along the backbone only.

4.2 Statistical properties

4.2.1 N finite

If the local waiting time pdf $\phi_0(t)$ has finite moments, its Laplace transform reads [3], $\hat{\phi}_0(s) \simeq 1 - s\bar{t}$, in the large time limit $s \to 0$, where $\bar{t}$ is the local mean waiting time at each node.[1] Taking the limit $s \to 0$ in Eq. (4.23), we obtain the waiting time pdf for the effective backbone dynamics,

$$\hat{\psi}(s) \simeq (1 + s \langle t \rangle)^{-1}. \tag{4.24}$$

The mean waiting time $\langle t \rangle$ is given by

$$\langle t \rangle = \int_0^\infty t\psi(t)dt = \frac{\bar{t}}{2q-1} \left[2(1-q)^{1-N}q^N + 4q - 3 \right]. \tag{4.25}$$

In Fig. 4.2, we plot the effective mean waiting time at a node of the backbone dynamics versus N and q. The mean waiting time $\langle t \rangle$ is a monotonically increasing function of both q and N. If the random walk inside the branches is isotropic, $q = 1/2$, one obtains by L'Hopital's rule from Eq. (4.25)

$$\lim_{q \to 1/2} \langle t \rangle = (1 + 2N)\,\bar{t}. \tag{4.26}$$

To determine the diffusion coefficient D for diffusion through the comb, we first calculate the MSD. Performing the Fourier–Laplace transformation of Eq. (4.1), we obtain

$$\hat{P}(k,s) = \frac{1 - \hat{\psi}(s)}{s[1 - \tilde{\Phi}(k)\hat{\psi}(s)]}. \tag{4.27}$$

The MSD in Laplace space reads

$$\langle x^2(s) \rangle = -\lim_{k \to 0} \frac{d^2 P(k,s)}{dk^2}. \tag{4.28}$$

As mentioned above, we assume that the motion on the backbone is unbiased and the walker jumps only to the nearest neighbors. This implies that

---

[1]To some extent, it suffices if only the first two moments are finite: $\bar{t} = \int_0^\infty t\phi_0(t)dt$ and $\bar{t}^2 = \int_0^\infty t^2\phi_0(t)dt$. If one introduces the pdf $\Phi_0(t) = \int_t^\infty \phi_0(\tau)d\tau$ of waiting times larger than t for the local pdf $\phi_0(t)$, then from the first moment we have $1 = \int_0^\infty \phi_0(t)dt = -\frac{1}{\bar{t}} \int_0^\infty t\Phi_0'(t)dt = \frac{1}{\bar{t}} \int_0^\infty \Phi_0(t)dt$. The second moment yields $\frac{2\bar{t}}{\bar{t}^2} \int_0^\infty t\Phi_0(t)dt = \int_0^\infty \Phi_0(t)dt$. From these two relations, it follows that if $\phi_0(t) = A\Phi_0(t)$, which yields the differential equation $\dot{\phi}_0 = -A\phi_0$, then $\phi_0(t) = Ae^{-At}$, and one easily obtains $A = 1/\bar{t}$ and $\bar{t}^2 = 2(\bar{t})^2$. The Laplace transformation yields $\hat{\phi}_0(s) = (1 + s\bar{t})^{-1} \approx 1 - s\bar{t}$ for $s\bar{t} \ll 1$.

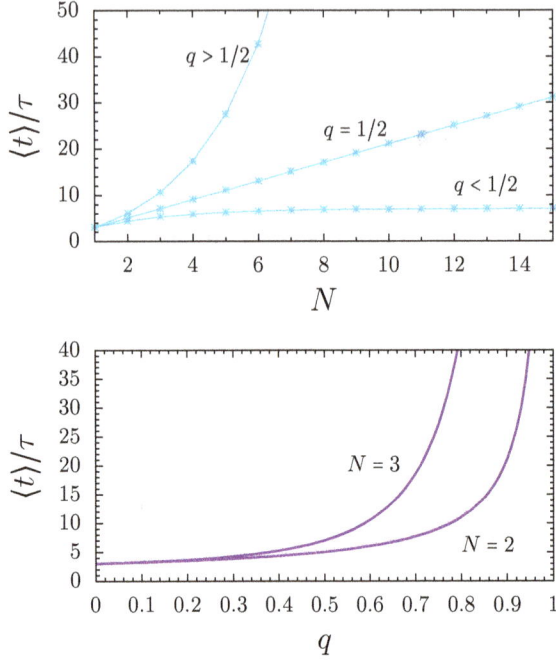

Fig. 4.2 Dimensionless mean waiting time of the effective backbone dynamics. Reprinted figure with permission from V. Méndez, A. Iomin, D. Campos, and W. Horsthemke, Phys. Rev. E, 92, 062112 (2015). Copyright 2015 by the American Physical Society.

the kernel $\Phi(x)$ is $\Phi(x) = \delta(x-a)/2 + \delta(x+a)/2$. Then we obtain from Eq. (4.28),

$$\langle x^2(s) \rangle = \frac{a^2}{s\left[\hat{\psi}(s)^{-1} - 1\right]}. \tag{4.29}$$

If the waiting time pdf $\psi(t)$ possesses a finite first moment, Eq. (4.24) implies that the MSD along the backbone corresponds to normal diffusion,

$$\langle x^2(t) \rangle = 2Dt. \tag{4.30}$$

The diffusion coefficient is given by

$$D = \frac{a^2}{2\langle t \rangle} = \frac{a^2}{2\bar{t}} \frac{2q-1}{2(1-q)^{1-N}q^N + 4q - 3}. \tag{4.31}$$

Note that in the limit of an isotropic random walk, i.e., when $q = 1/2$, D behaves like N^{-1} for large N by virtue of Eqs. (4.26) and (4.31). As Fig. 4.2 demonstrates, $\langle t \rangle$ increases monotonically with N for $q < 1/2$ and saturates at $(4q-3)/(2q-1)$ for $N \to \infty$. Consequently, the mean waiting time $\langle t \rangle$ is finite for $N \to \infty$; the overall diffusion along the backbone is normal. However, for $q \geq 1/2$, the mean waiting time $\langle t \rangle$ increases without bound as N increases, and anomalous transport is expected for $N \to \infty$. However for finite N, the transport is diffusive regardless q and the specific form of $\phi_0(\tau)$, as long as it has finite moments.

We consider now a local waiting time pdf with the large time limit $\phi_0(\tau) \sim \tau^{-1-\gamma}$, with Laplace transform $\hat{\phi}_0(s) \simeq 1 - (s\tau_0)^\gamma$ and $0 < \gamma < 1$, which does not possess finite moments. Here τ_0 is a parameter with units of time. In this case, the waiting time pdf for the backbone dynamics is obtained by simply replacing $s\bar{t}$ with $(s\tau_0)^\gamma$, i.e,

$$\phi(s) \simeq [1 + (s\tau_0)^\gamma \langle t \rangle / \tau_0]^{-1}. \tag{4.32}$$

Substituting this result into Eq. (4.29), we find

$$\langle x^2(t) \rangle = \frac{a^2 \tau_0}{\langle t \rangle} \frac{(t/\tau_0)^\gamma}{\Gamma(1+\gamma)}, \tag{4.33}$$

for large t, where $\langle t \rangle$ is given by Eq. (4.25), with τ_0 instead of $\bar{t}$. If the waiting time pdf $\phi_0(\tau)$ at each node of the comb has a power-law tail, then the overall transport along the backbone is anomalous.

4.2.2 $N \to \infty$

If the number of nodes of the branches goes to infinity, the mean time spent by the walker visiting a branch increases monotonically, according Eq. (4.25). If the diffusion coefficient tends asymptotically to a constant, which is the case for $q < 1/2$, the diffusive scaling will saturate at

$$D = a^2(2q-1)/[2\bar{t}(4q-3)], \tag{4.34}$$

according to Eq. (4.31). For $q \geq 1/2$, the limit $N \to \infty$ leads to $D \to 0$ and we expect a different scaling. For $N \to \infty$, the quotient $(\lambda_-/\lambda_+)^N \to 0$ and also $\lambda_N \to 0$. We obtain from Eq. (4.21a),

$$\Lambda(q, \hat{\phi}_0(s)) = \frac{2q\hat{\phi}_0(s)}{1 + \sqrt{1 - 4q(1-q)\hat{\phi}_0^2(s)}} \equiv \frac{2q\hat{\phi}_0(s)}{1 + g(q)}, \tag{4.35}$$

where we define $g(q) \equiv \sqrt{1 - 4q(1-q)\hat{\phi}_0^2(s)}$ for convenience. Equation (4.23) for the Laplace transform of the effective waiting time pdf reduces to

$$\hat{\psi}(s) = \frac{\hat{\phi}_0(s)\left[1 + g(q) - 2q(1-q)\hat{\phi}_0^2(s)\right]}{2 - (1 + 3q - 4q^2)\hat{\phi}_0^2(s) + [2 - (1-q)\hat{\phi}_0^2(s)]g(q)}. \tag{4.36}$$

We take the limit $s \to 0$ and consider the case where $\phi_0(\tau)$ has finite moments. Then $\hat{\phi}_0(s) \simeq 1 - s\bar{t}$, as $s \to 0$.[2] The square root $g(q)$ in Eq. (4.36) reads

$$
g(q) \simeq \begin{cases} 1 - 2q - \dfrac{4q(1-q)}{2q-1}s\bar{t}, & q < 1/2, \\[3mm] \sqrt{2s\bar{t}} - \dfrac{\sqrt{2}}{4}(s\bar{t})^{3/2}, & q = 1/2, \\[3mm] -1 + 2q + \dfrac{4q(1-q)}{2q-1}s\bar{t}, & q > 1/2, \end{cases} \tag{4.37}
$$

and Eq. (4.36) implies that the waiting time pdf of the backbone dynamics is given by

$$
\hat{\psi}(s) \simeq \begin{cases} \left(1 + \dfrac{4q-3}{2q-1}s\bar{t}\right)^{-1}, & q < 1/2, \\[3mm] \left(1 + \sqrt{2s\bar{t}}\right)^{-1}, & q = 1/2, \\[3mm] \left(\dfrac{3q-1}{q} + \dfrac{4q^2-3q+1}{(2q-1)q}s\bar{t}\right)^{-1}, & q > 1/2. \end{cases} \tag{4.38}
$$

Substituting Eq. (4.38) into Eq. (4.29), we find for large t,

$$
\langle x^2(t)\rangle = \begin{cases} a^2\dfrac{2q-1}{4q-3}\cdot\dfrac{t}{\bar{t}}, & q < 1/2, \\[3mm] a^2\sqrt{\dfrac{2t}{\pi\bar{t}}}, & q = 1/2, \\[3mm] a^2\dfrac{q}{2q-1}\left(1 - e^{-\kappa t}\right), & q > 1/2, \end{cases} \tag{4.39}
$$

where the rate of saturation is

$$
\kappa = \frac{(2q-1)^2}{(4q^2-3q+1)\bar{t}}. \tag{4.40}
$$

In Fig. 4.3 we compare these results with numerical simulations for $N = 10^4$. For $q = 1/2$, we obtain the well known result of subdiffusive transport with the MSD $\sim \sqrt{t}$. However, for $q \neq 1/2$, the side branches experience advection, and the transport is remarkably different. Namely, for $q > 1/2$ the advection is away from the backbone along the branches, $y \to \pm\infty$. The walker is effectively trapped inside the branches, and stochastic localization (diffusion failure) occurs, $\langle x^2(\infty)\rangle < \infty$, [84]. For $q < 1/2$, the advection is towards the backbone. It enhances the backbone dynamics

---

[2]Note that the latter condition is not necessary, if $\bar{t}$ is small enough, such that $s\bar{t} \ll 1$.

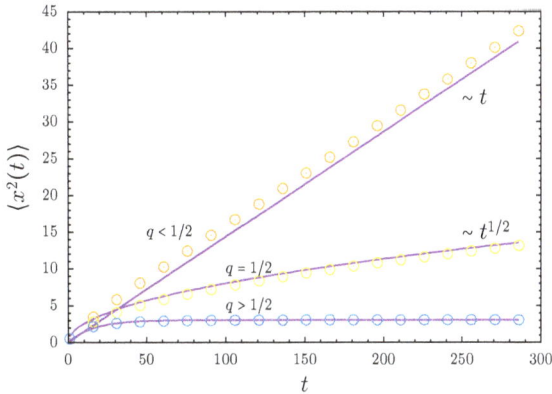

Fig. 4.3   MSD for three values of q, displaying the three different transport regimes. Solid curves correspond to the results given by Eq. (4.39) for $q = 0.4$, $q = 0.5$, and $q = 0.6$. Symbols correspond to the results of numerical simulations with $N = 10^4$ and $a = 1$. Reprinted figure with permission from V. Méndez, A. Iomin, D. Campos, and W. Horsthemke, Phys. Rev. E, 92, 062112 (2015). Copyright 2015 by the American Physical Society.

and normal diffusion takes place. More details related to various examples are considered in Ref. [85].

Of course, the limit $N \to \infty$ cannot be attained in a strict sense for real systems. However, transport on a comb structure will display the behavior discussed above for N sufficiently large. Note that the numerical results in Fig. 4.3 were obtained for $N = 10^4$. On the other hand, we expect that Eq. (4.39) holds experimentally only up to a large finite time, namely as long as the walker does not experience the finite size of the branches. This experimental situation is considered in Chapter 9 for the actin polymerization in a comb micrograph.

4.2.3   *N random*

The waiting time pdf, obtained in Eq. (4.23) is the general result, valid for all possible realizations of the branch length N. An interesting case is a comb with varying branch lengths, where N is a random number,[3] such

---

[3]This problem was addressed in many studies of the transport properties of networks. Investigations specifically related to the comb with random branches can be found, for example, in Refs. [86, 87, 14, 88, 89, 35, 36].

that the branch length $L = Na$ is distributed according to the power law

$$\rho(L) = \frac{a^{-1-\gamma}}{\zeta(\gamma+1)}(N)^{-1-\gamma}, \tag{4.41}$$

where $\zeta(\gamma+1)$ is the Riemann zeta function. An analytical estimate of the average waiting time pdf

$$\langle \psi(t) \rangle = \sum_{N=1}^{\infty} \rho(L)\psi(t, N), \tag{4.42}$$

where $\psi(t, N) \equiv \psi(t)$, defined in Eq. (4.23), is a hard to obtain, because Eq. (4.23) is a very complicated function of N. Another obstacle is that the limits $s \to 0$ and $N \to \infty$ cannot be considered straightforwardly, since the limiting case at the boundaries at $y = Na$ is not well defined anymore. To overcome this difficulty we use the following assumptions, suggested in Ref. [36], which are suitable for $q = 1/2$.

The backbone dynamics corresponds to normal diffusion for finite branches according to Eq. (4.24). However, the condition[4] $\lambda_N = 0$, which follows from $N \to \infty$, means that a walker does not reach the boundary at $y = Na$. Therefore, it can be used for finite branches on the time scale when the branched MSD is much less then Na, as well. This is the reason why the limit $s \to 0$ is not valid, since it depends on the averaging procedure. For example, for the unbiased motion with $q = 1/2$, this condition reads $N > \sqrt{t/\bar{t}}$, where diffusion along the branches has the diffusion coefficient $D_y = a^2/\bar{t}$. Therefore, for the time scale $t < \frac{L^2}{D_y} = N^2\bar{t}$, the backbone dynamics is subdiffusive and described by Eqs. (4.38) and (4.39). On this time scale, the waiting time pdf corresponds to subdiffusion and is independent of N. Note that Eq. (4.37) for $g(q)$ is valid as well, see the comment in Footnote 2. Now we perform the averaging procedure, which is completely independent of N and the limit $N \to \infty$. In this case, the averaging procedure reduces to the multiplication of the $\psi(t)$ by the factor $\frac{a^{-1-\gamma}}{\zeta(\gamma+1)} \sum_{N=N_0}^{\infty} (N)^{-1-\gamma}$, where $N_0 = \sqrt{t/\bar{t}}$.

Considering this large scale of the branches for the averaging procedure, it is convenient to treat the branch length L as a continuous variable. Therefore, the discrete distribution in Eq. (4.41) is replaced by a continuous distribution for $\rho(L)$ that reads

$$\rho(L) = \gamma L^{-1-\gamma}, \quad L > 1, \quad 0 < \gamma < 1, \tag{4.43}$$

such that

$$\int_{1}^{\infty} \rho(L)dL = 1, \quad \text{and} \quad \langle L \rangle = \int_{1}^{\infty} L\rho(L)dL = \infty. \tag{4.44}$$

---

[4]See Eq. (4.21b).

The averaging procedure reduces to the integration

$$\langle \psi(t) \rangle_L = \psi(t) \int_{a\sqrt{t/\bar{t}}}^{\infty} \rho(L) dL = \left(\frac{at}{\bar{t}} \right)^{-\frac{\gamma}{2}} \psi(t) \sim t^{-\frac{3+\gamma}{2}}. \tag{4.45}$$

This, eventually, leads to the MSD of the backbone subdiffusion for $q = 1/2$, which reads

$$\langle x^2(t) \rangle \sim t^{\frac{1+\gamma}{2}}. \tag{4.46}$$

An alternative way to describe combs with finite branches is suggested in Chapter 5. It is based on the eigenvalue problem in fingers, which leads to the same result as Eq. (4.46), and for random N it has been described in Ref. [35], as well.

Chapter 5

Langevin formalism

5.1 Langevin equation

The general random dynamics of a particle with unit mass along a comb structure is described by dynamical equations of motion with a random force $\mathbf{F}$ and friction γ, i.e., a system of Langevin equations:

$$\ddot{X}(t) - \gamma\dot{x}(t) = F_x(X(t), Y(t), t), \quad \ddot{Y}(t) - \gamma\dot{y}(t) = F_y(X(t), Y(t), t). \quad (5.1)$$

Specifically, we assume that the vector force has the following form,

$$\mathbf{F} = (F_x, F_y) = (\gamma C(Y(t))\xi_x(t), \gamma\xi_y(t)). \quad (5.2)$$

Here $\xi_x(t)$ and $\xi_y(t)$ are two uncorrelated Gaussian white noises whose probability distribution functions are given by the functional form

$$[d\rho(\xi_i)] = \exp\left[-\frac{1}{4D_i}\int \xi_i^2(t)dt\right][d\xi_i(t)], \quad (5.3)$$

where $[d\xi_i(t)] = \prod_{\tau=0}^{t} d\xi_i(\tau)/\sqrt{4\pi D_i/d\tau}$. Since the noises $\boldsymbol{\xi} = (\xi_x, \xi_y)$ are Gaussian, they are also completely described by their means and correlation functions:

$$\langle \xi_x(t)\xi_x(t')\rangle = 2D_x\delta(t-t'), \quad (5.4)$$
$$\langle \xi_y(t)\xi_y(t')\rangle = 2D_y\delta(t-t'), \quad (5.5)$$
$$\langle \xi_x(t)\xi_y(t')\rangle = \langle \xi_x(t)\rangle\langle \xi_y(t')\rangle = \langle \xi_x(t)\rangle = \langle \xi_y(t)\rangle = 0. \quad (5.6)$$

Here $\langle\ldots\rangle$ denotes averaging over the Gaussian distribution (5.3) of the random field $\boldsymbol{\xi} = (\xi_x, \xi_y)$. In the overdamped case, the Langevin equations read,

$$\dot{X} = C(Y)\xi_x(t), \quad \dot{Y} = \xi_y(t), \quad (5.7)$$

where the $-$ sign has been absorbed in the random fields. The solution of these equations can be written in the form of functionals,

$$X(t) = x[\xi_x(t')] = X_0 + \int_0^t C[y(t')]\xi_x(t')dt', \tag{5.8}$$

$$Y(t) = y[\xi_y(t')] = Y_0 + \int_0^t \xi_y(t')dt', \tag{5.9}$$

where $X_0 = X(t=0)$ and $Y_0 = Y(t=0)$ are the initial conditions.

5.1.1 *From Langevin equations to the Fokker–Planck equation*

The evolution of the system, described by the Langevin equations (5.7), corresponds to a stochastic process $(X(t), Y(t))$ given by Eqs. (5.8) and (5.9). The latter can be described by a pdf $P(x, y, t)$, which is defined formally as follows [90],

$$P(x, y, t) = \langle\langle \delta[X(t) - x]\delta[Y(t) - y]\rangle\rangle, \tag{5.10}$$

where averaging $\langle\langle \cdots \rangle\rangle$ is performed over the random field $\boldsymbol{\xi}$. Here x, y, and t are independent variables, and the expectation value $P(x, y, t)$ in (5.10) is the pdf to find a particle at time t at the position (x, y). Differentiating $P(x, y, t)$ with respect to time, we obtain

$$\frac{\partial P(x, y, t)}{\partial t} = \langle\langle \frac{\partial}{\partial X(t)} \delta[X(t) - x)]\dot{X}(t)\delta[Y(t) - y]\rangle\rangle$$

$$+ \langle\langle \frac{\partial}{\partial Y(t)} \delta[X(t) - x]\dot{Y}(t)\delta[Y(t) - y]\rangle\rangle. \tag{5.11}$$

We take into account the Langevin equations (5.7) and use the property of the δ-function, which allows us to change the derivatives $\partial/\partial X(t)$ and $\partial/\partial Y(t)$ to $\partial/\partial x$ and $\partial/\partial y$ and $C(Y(t))$ to $C(y)$ and take them outside of the brackets. In this way we obtain from Eq. (5.11),

$$\frac{\partial P(x, y, t)}{\partial t} = C(y)\frac{\partial}{\partial x}\langle\langle \xi_x(t)\delta[X(t) - x]\delta[Y(t) - y]\rangle\rangle$$

$$+ \frac{\partial}{\partial y}\langle\langle \delta[X(t) - x]\xi_y(t)\delta[Y(t) - y]\rangle\rangle, \tag{5.12}$$

which is however not in closed form. To obtain a closed version, we need to evaluate the averages $\langle\langle \delta[X(t) - x]\boldsymbol{\xi}(t)\delta[Y(t) - y]\rangle\rangle$. To this end we use the following auxiliary expression

$$\frac{2}{D}\xi(t)\exp\left[-\frac{1}{D}\int \xi^2(\tau)d\tau\right] = -\frac{\delta}{\delta\xi(t)}\exp\left[-\frac{1}{D}\int \xi^2(\tau)d\tau\right]. \tag{5.13}$$

Functional Gaussian integration by parts yields

$$\langle \Phi[\xi]\xi(t) \rangle = \int \xi(t)\Phi[\xi] \exp\left[-\frac{1}{D}\int \xi^2(\tau)d\tau\right][d\xi(t)]$$

$$= \int \exp\left[-\frac{1}{D}\int \xi^2(t)dt\right]\frac{\delta\Phi[\xi]}{\delta\xi(t)} = \left\langle \frac{\delta\Phi[\xi]}{\delta\xi(t)} \right\rangle. \qquad (5.14)$$

A simple generalization is given by the Furutsu–Novikov formula [91, 92],

$$\langle \xi(t)\Phi[\xi(\tau)] \rangle = \int \langle \xi(t)\xi(t') \rangle \left\langle \frac{\delta\Phi[\xi(\tau)]}{\delta\xi(t')} \right\rangle dt'. \qquad (5.15)$$

In the case of δ-correlated noise, Eq. (5.15) reduces to Eq. (5.14). We apply the Furutsu–Novikov formula (5.15) to Eq. (5.12) and obtain the following chain of calculations for $\xi_x(t)$,

$$\langle\langle \xi_x(t)\delta[X(\tau) - x]\delta[Y(\tau) - y]\rangle\rangle$$

$$= \int \langle \xi_x(t)\xi_x(t') \rangle \left\langle\left\langle \frac{\delta}{\delta\xi_x(t')}\delta[X(\tau) - x]\delta[Y(\tau) - y]\right\rangle\right\rangle$$

$$= 2D_x \left\langle\left\langle \frac{\partial}{\partial X(t)}\delta[X(\tau) - x]\frac{\delta X(\tau)}{\delta\xi_x(t)}\delta[Y(\tau) - y]\right\rangle\right\rangle. \qquad (5.16)$$

Analogously, we obtain for $\xi_y(t)$,

$$\langle\langle \xi_y(t)\delta[X(\tau) - x]\delta[Y(\tau) - y]\rangle\rangle$$

$$= 2D_y \left\langle\left\langle \delta[X(\tau) - x]\frac{\partial}{\partial Y(t)}\delta[Y(\tau) - y]\frac{\delta Y(\tau)}{\delta\xi_y(t)}\right\rangle\right\rangle. \qquad (5.17)$$

For the functional differentiation $\delta x(\tau)/\delta\xi_x(t)$ and $\delta y(\tau)/\delta\xi_y(t)$ we take into account the principle of causality which yields nonzero results only for $\tau \geq t$. We obtain from the functional solution (5.9)

$$\frac{\delta y(\tau)}{\delta\xi_y(t)} = \lim_{\Delta t \to 0}\int_{t-\frac{\Delta t}{2}}^{t}\frac{\delta\xi_y(t')}{\delta\xi_y(t)}dt' = \frac{1}{2}. \qquad (5.18)$$

In the same way, using the midpoint rule and the principle of causality, we obtain from the solution (5.8)

$$\frac{\delta x(\tau)}{\delta\xi_x(t)} = \lim_{\Delta t \to 0}\int_{t-\frac{\Delta t}{2}}^{t} C(Y(t'))\frac{\delta\xi_x(t')}{\delta\xi_x(t)}dt'$$

$$= C[Y(t)]\lim_{\Delta t \to 0}\int_{t-\frac{\Delta t}{2}}^{t}\frac{\delta\xi_x(t')}{\delta\xi_x(t)}dt' = \frac{C(Y(t))}{2}, \qquad (5.19)$$

where $C[Y(t)]$ is considered to be an implicit function of t. We substitute these derivatives into Eqs. (5.16) and (5.17) and the latter expressions into Eq. (5.12) to obtain the Fokker–Planck equation,

$$\frac{\partial P(x,y,t)}{\partial t} = D_x C^2(y) \frac{\partial^2 P(x,y,t)}{\partial x^2} + D_y \frac{\partial^2 P(x,y,t)}{\partial y^2}. \tag{5.20}$$

The comb model corresponds to $C^2(y) = \delta(y)$. While the use of the Dirac δ-function poses no problem in the Fokker–Planck equation, since the generalized function appears together with the function $P(x,y,t)$, its meaning in the Langevin equation (5.7), in the form $C(y) = \sqrt{\delta(y)}$, requires some care. We interpret the Langevin equation by starting with the regularized δ-function $C(y) = \exp\left(-y^2/2\epsilon\right)/\sqrt[4]{\epsilon\pi}$. In other words, we consider the dynamics inside a narrow strip of width ϵ for the Langevin equation and then take the limit $\epsilon \to 0$. Various aspects of the relation between the Langevin equation and the comb model have been considered in Refs. [93–95].

5.1.2   *Auxiliary: Functional derivative*

To establish a kinetic description of combs, we use functional (variational) derivatives. For the convenience of the reader, we briefly discuss functional differentiation here. For more details, we recommend the Refs. [96, 97, 90, 98], which provide a mathematical foundation of functional theory [96] and its application to the theory of stochastic differential equations [97, 90, 98].

A quantity F is a functional of the function $\xi(t)$, if there is a rule that gives rise to a unique value $F[\xi(t)]$ for every function $\xi(t)$, defined on the interval $[a, b]$.

Example 5.1. For example, a linear functional can be defined as follows

$$F[\xi] = \int_{t_a}^{t_b} A(t)\xi(t)dt, \tag{5.21}$$

where $A(t)$ is a continuous function. Functionals can be nonlinear as well. For example, a bilinear functional reads

$$F[\xi] = \int_{t_a}^{t_b} \int_{t_a}^{t_b} B(t_1, t_2)\xi(t_1)\xi(t_2)dt_1 dt_2. \tag{5.22}$$

Another form of nonlinear functionals is given by

$$F[\xi] = f\Big(\Phi[\xi(t)]\Big), \tag{5.23}$$

where $f(\Phi)$ is a known well-defined function, while $\Phi[\xi(t)]$ itself is a functional. $\square$

In our case, the solution $Y(t)$ in Eq. (5.9) is a linear functional with respect to the field $\xi_y(t)$, where $A(t) = 1$. Solution $X(t)$ is also a linear functional with respect to the field $\xi_x(t)$, where $A(t) = C(y(t))$. However, with respect to the field ξ_y, the solution $X(t)$ is a nonlinear functional of the form (5.23).

The variation of the functional $F[\xi(t)]$ is the difference

$$\delta F[\xi] = F[\xi(t) + \delta\xi(t)] - F[\xi(t)], \qquad (5.24)$$

where $\delta\xi(t)$ is nonzero in the vicinity Δt of a point $t \in [t_a, t_b]$. Note that $\delta F[\xi]$ is a linear functional with respect to the variation $\delta\xi(t)$,

$$\delta F[\xi] = \int_{t_a}^{t_b} F'[\xi(t), t]\delta\xi(t)dt. \qquad (5.25)$$

Here, in complete analogy with an ordinary derivative, $F'[\xi(t), t]$ is the functional derivative at the point t,

$$F'[\xi(t), t] \equiv \frac{\delta F[\xi]}{\delta\xi(t)} = \lim_{\Delta t \to 0} \frac{\delta F[\xi]}{\int_{\Delta t} \delta\xi(\tau)d\tau}. \qquad (5.26)$$

It is instructive to determine the functional derivatives of the functionals given in Example 5.1. Calculating the functional derivative of the linear functional (5.21) we have

$$\delta F[\xi] = F[\xi + \delta\xi] - F[\xi] = \int_{t_a}^{t_b} A(t)\delta\xi(t)dt$$

$$= \int_{\Delta t} A(t)\delta\xi(t)dt = \int_{t-\frac{\Delta t}{2}}^{t+\frac{\Delta t}{2}} A(\tau)\delta\xi(\tau)d\tau = A(t)\int_{\Delta t}\delta\xi(t)dt. \quad (5.27)$$

According the definition (5.26), the functional derivative of the linear functional (5.21) is $\delta F[\xi]/\delta\xi(t) = A(t)$. An important *corollary* is the derivative at the bounds of integration,

$$\frac{\delta F[\xi]}{\delta\xi(t_b)} = \frac{1}{2}A(t_b). \qquad (5.28)$$

A second important corollary is

$$\frac{\delta\xi(\tau)}{\delta\xi(t)} = \delta(t - \tau). \qquad (5.29)$$

Functional differentiation of the bilinear functional (5.22) yields

$$\frac{\delta F[\xi]}{\delta\xi(t)} = \int_{t_a}^{t_b} [B(t, t') + B(t', t)]\xi(t')dt'. \qquad (5.30)$$

Consequently, the second variational derivative is given by

$$\frac{\delta^2 F[\xi]}{\delta\xi(t_1)\delta\xi(t_2)} = \frac{\delta}{\delta\xi(t_2)}\left[\frac{\delta F[\xi]}{\delta\xi(t_1)}\right] = B(t_1, t_2) + B(t_2, t_1). \tag{5.31}$$

Differentiation of the nonlinear functional in Eq. (5.23) can be carried out by means of the standard chain rule.

$$\frac{\delta F[\xi]}{\delta\xi(t)} = \frac{\partial f(\Phi)}{\partial\Phi}\frac{\delta\Phi[\xi]}{\delta\xi(t)}. \tag{5.32}$$

We can generalize the chain rule to a composite functional of the form $F_1[\xi(t)] = F\big[\Phi[\zeta(t');t]\big]$. Such expressions appear if instead of the function $\xi(t)$ a new function $\zeta(t)$ is introduced by means a new functional $\xi(t) = \Phi[\zeta(t');t]$. The chain rule yields[1]

$$\frac{\delta F_1[\xi(\tau)]}{\delta\zeta(t)} = \int_{t_a}^{t_b} \frac{\delta F[\xi(\tau)]}{\delta\xi(t')}\frac{\delta\Phi(\zeta(t'');t')}{\delta\zeta(t)}dt'. \tag{5.33}$$

5.2   Mean squared displacement

To determine the type of transport behavior displayed by the ramified structure, we need to determine the mean squared displacement (MSD) of the motion along the backbone, $X(t)$, which is determined by the functional solutions (5.8) and (5.9). We begin by averaging over the noise ξ_x. Taking $X_0 = 0$, we obtain

$$\langle X^2(t)\rangle = \int_0^t\int_0^t C[Y(t')]C[Y(t'')]\langle\xi_x(t')\xi_x(t'')\rangle dt'dt'', \tag{5.34}$$

which yields, taking into account Eq. (5.4),

$$\langle X^2(t)\rangle = 2D_x\int_0^t C^2[Y(t')]dt'. \tag{5.35}$$

Next we average over the random field ξ_y. This can be done by using the marginal pdf

$$P_Y(y,t) = \int P(x,y,t)dx = \langle\delta[Y(t) - y]\rangle. \tag{5.36}$$

We obtain for the MSD

$$\langle\langle X^2(t)\rangle\rangle = 2D_x\int_{\Omega_Y} C^2(y)\int_0^t P_Y(y,t')dt'. \tag{5.37}$$

---

[1]This explanation is somewhat simplistic. A mathematically rigorous derivation can be found in Ref. [99].

Here Ω_Y is the range of $Y(t)$: $Y \in \Omega_Y$. The simplest case occurs if the structure is actually not ramified at all, i.e., the particles move in the x-y-plane. The dynamics of $X(t)$ and $Y(t)$ are independent, i.e., $C[Y(t)] = C = \text{const}$, and we obtain

$$\langle\langle X^2(t) \rangle\rangle = 2D_x C^2 t. \tag{5.38}$$

In other words, the motion projected onto the x-axis corresponds to normal diffusion. This is the expected result, since the dynamics of $X(t)$ is independent of $Y(t)$ and is driven by white noise.

A more interesting situation occurs for a comb-like structure, where $C^2(y) = \delta(y)$. In this case, the MSD in Eq. (5.37) reads

$$\langle\langle X^2(t) \rangle\rangle = 2D_x \int_0^t P_Y(y = 0, t')dt'. \tag{5.39}$$

It is useful to write the above result in Laplace space,

$$\langle\langle X^2(s) \rangle\rangle = 2D_x \frac{\hat{P}_Y(y = 0, s)}{s}. \tag{5.40}$$

5.2.1 Anomalous diffusion along the branches

Equation (5.40) shows that transport along the backbone is determined by the random motion along the fingers (teeth). Changing the finger dynamics, we obtain a variety of backbone transport regimes. To deal with the dynamics along the fingers, we will consider known propagators for $P_Y(y = 0, t)$. If the fingers have infinite lengths, then $\Omega_Y = \mathbb{R}$, and it is convenient to work in Fourier space. Given the definition of the inverse Fourier transform,

$$P_Y(y, t) = (2\pi)^{-1} \int_{-\infty}^{\infty} P_Y(k, t) \exp(-iky)dk, \tag{5.41}$$

we obtain P_Y at the backbone as

$$P_Y(y = 0, t) = (2\pi)^{-1} \int_{-\infty}^{\infty} P_Y(k, t)dk. \tag{5.42}$$

The MSD in Eq. (5.40) reads in Laplace space

$$\langle\langle \hat{X}^2(s) \rangle\rangle = \frac{D_x}{\pi s} \int_{-\infty}^{\infty} \hat{P}_Y(k, s)dk. \tag{5.43}$$

Below we consider different dynamics along the branches and obtain the MSD of the overall structure in each case.

5.2.1.1 Subdiffusion in fingers

If the motion in the y-direction corresponds to a continuous-time random walk with a heavy-tailed waiting time pdf, $\psi(t) \sim t^{-1-\alpha}$ with $0 < \alpha < 1$, then the motion along the teeth is subdiffusive. In this case, the marginal pdf $P_Y(y,t)$ in Fourier–Laplace space in the diffusive limit $(k,s) \to 0$ is given by the Montroll–Weiss equation [5],

$$\hat{P}_Y(k,s) = \frac{1}{s + D_y k^2 s^{1-\alpha}}. \tag{5.44}$$

This expression is also discussed in Refs. [3, 4], see also Sec. 1.3.3. We substitute this expression into Eq. (5.43) and carry out the inverse Laplace–Fourier transform:

$$\langle\langle X^2(t)\rangle\rangle = \frac{D_x t^{1-\alpha/2}}{D_y \Gamma(2-\alpha/2)}. \tag{5.45}$$

In other words, subdiffusion in the y-direction with anomalous transport exponent α gives rise to subdiffusive transport through the ramified structure along the backbone with anomalous transport exponent $1 - \alpha/2$. This result agrees with the macroscopic result obtained assuming subdiffusive transport along the branches and normal diffusion along the backbone in Ref. [16], see also Chapter 3. As the motion in the y-direction approaches normal diffusive behavior with $\alpha = 1$, the MSD approaches the behavior $\langle\langle X^2(t)\rangle\rangle \sim \sqrt{t}$, which is well-known from the study of simple random walks on combs, considered in the Introduction.

5.2.1.2 Lévy flights in fingers

If the motion in the y-direction corresponds to a continuous-time random walk with a heavy-tailed jump length pdf, $\lambda(y) \sim |y|^{-1-\mu}$ with $1 < \mu < 2$, i.e., $Y(t)$ is a Lévy flight, then the motion is superdiffusive. In this case, the marginal pdf $P_Y(y,t)$ corresponds to the Lévy distribution [3, 4] and reads in Fourier–Laplace space

$$\hat{P}_Y(k,s) = \frac{1}{s + D_y |k|^\mu}. \tag{5.46}$$

We substitute this expression into Eq. (5.43) and carry out the inverse Fourier and Laplace transformations:

$$\langle\langle X^2(t)\rangle\rangle = \frac{2D_x}{\pi\mu D_y^{1/\mu}\sin(\pi/\mu)} \cdot \frac{t^{1-1/\mu}}{\Gamma(2-1/\mu)}. \tag{5.47}$$

Interestingly, superdiffusive motion in the y-direction also gives rise to subdiffusive transport through the ramified structure, i.e., along the backbone, with the anomalous transport exponent $1 - 1/\mu < 1/2$.

5.2.1.3 *Diffusion in finite-length fingers*

If the range Ω_Y of $Y(t)$ corresponds to a finite interval, it is convenient to work with Eq. (5.39) or Eq. (5.40). These expressions are particularly convenient if the dynamics on the secondary branches is described by a diffusion equation. As an example, we consider normal diffusion along one-dimensional branches in the y-direction of length $2L$,

$$\partial_t P_Y(y,t) = \partial_y^2 P_Y(y,t), \tag{5.48}$$

with reflecting boundary conditions $\partial_y P_Y(y,t)|_{y=\pm L} = 0$ and the initial condition $P_Y(y, t = 0) = \delta(y)$. Here $\partial_y \equiv \partial/\partial y$. The pdf $P_Y(y,t)$ is given by the Fourier series

$$P_Y(y,t) = \frac{1}{2L} \sum_{n=-\infty}^{\infty} \exp\left(-\frac{\pi^2 n^2}{L^2}t\right) \exp\left(-i\frac{\pi n y}{L}\right). \tag{5.49}$$

At $t = 0$, the Fourier sum reads, according the Poisson summation formula,

$$\frac{1}{2L} \sum_{n=-\infty}^{\infty} \exp\left(-i\frac{\pi n y}{L}\right) = \sum_{m} \delta(y - 2mL). \tag{5.50}$$

Since $y \in [-L, L]$, one obtains simply the initial condition at $m = 0$. Consequently, Eqs. (5.39) and (5.49) yield

$$\langle\langle X^2(t)\rangle\rangle \to \frac{D_x}{L}t \quad \text{for} \quad t \to \infty, \tag{5.51}$$

i.e., transport through the comb is normal diffusion as expected.

We compare this result with the case where the diffusion along the fingers is anomalous. The equation for anomalous diffusion along one-dimensional branches in the y-direction of length $2L$ with generalized transport coefficient K_α is given by the FFPE

$$\partial_t P_Y(y,t) = \mathcal{D}^{1-\alpha} K_\alpha \partial_y^2 P_Y(y,t), \tag{5.52}$$

with $0 < \alpha < 1$ and with reflecting boundary conditions. Recall that the Riemann–Liouville fractional derivative $\mathcal{D}^\nu$ is defined in Sec. 2.5. In this case, the separation ansatz reads

$$P_Y(y,t) = \frac{1}{2L} \sum_{n=-\infty}^{\infty} T_n(t) \exp\left(-i\frac{\pi n y}{L}\right), \tag{5.53}$$

where $T_n(t) = T(e_n t)$ corresponds to the fractional relaxation equation

$$\frac{dT(e_n t)}{dt} = -e_n \mathcal{D}^{1-\alpha} T(e_n t), \tag{5.54}$$

with $e_n = K_\alpha \pi^2 n^2 / L^2$. Its solution is the Mittag–Leffler function [43]

$$T_n(t) = E_{\alpha,1}\left(-e_n t^\alpha\right) \equiv E_\alpha\left(-e_n t^\alpha\right) = \sum_{j=0}^{\infty} \frac{\left(-e_n t^\alpha\right)^j}{\Gamma(1+\alpha j)}, \qquad (5.55)$$

which corresponds to the initial condition choice $T_n(t=0) = 1$. For $\alpha = 1$ it reduces to the normal diffusion solution (5.49) with $K_1 = D_y$. Substituting ansatz (5.53) and Eq. (5.55) into Eq. (5.39), we obtain the MSD

$$\langle\langle X^2(t)\rangle\rangle = \frac{D_x t}{L} \sum_{n=-\infty}^{\infty} \sum_{j=0}^{\infty} \frac{\left(-e_n t^\alpha\right)^j}{\Gamma(2\alpha j)} \equiv \frac{t}{2L} \sum_{n=-\infty}^{\infty} E_{\alpha,2}\left(-e_n t^\alpha\right). \quad (5.56)$$

The Mittag–Leffler function behaves like a stretched-exponential for short times, [43, 3]

$$\exp\left[-e_n t^\alpha / \Gamma(1+\alpha)\right], \qquad (5.57)$$

and like a power-law for long times [43]),

$$E_{\alpha,\beta}\left(-e_n t^\alpha\right) \sim \frac{L^2}{\Gamma(\beta - \alpha) K_\alpha \pi^2 n^2 t^\alpha} \qquad \text{for } n \neq 0. \qquad (5.58)$$

Taking into account that $\sum_{n=1}^{\infty} 1/n^2 = \zeta(2) < \infty$, where $\zeta(\cdot)$ is the Riemann zeta function, we obtain that the main contribution to the transport is due to the term with $n = 0$. The asymptotic behavior of the MSD is given by

$$\langle\langle X^2(t)\rangle\rangle = \frac{D_x t}{L}, \qquad (5.59)$$

which coincides exactly with the normal diffusion result in Eq. (5.51). This implies that finite-length fingers give rise to the robust asymptotic result of normal diffusion along the backbone.

In light of the results obtained in this section we conclude that the overall MSD has different temporal exponents, i.e., different transport regimes are realized, depending on whether the length of the fingers is finite or infinite, even if the transport mechanism along the fingers is the same. Quantitatively, for infinite fingers, the MSD grows like $\sqrt{t}$, if diffusion along the fingers is normal, and MSD $\sim t^{1-\alpha/2}$ if the transport along the fingers is subdiffusive with the transport exponent α. For finite fingers, the situation changes dramatically. The boundary conditions lead to normal diffusion with the MSD $\sim t$, regardless of whether normal or anomalous diffusion takes place in the fingers. According to Eq. (5.47), fingers exert the strongest influence on the MSD if Lévy flights occur along the fingers. In particular, if $\mu = 1$, the transport along the backbone becomes ultraslow, and transport failure (stochastic localiztion) occurs for $0 < \mu < 1$, see Sec. 5.3.3.

5.3 Langevin dynamics along the branches: Characteristic functions

Characteristic functions of real-valued random variables completely define their probability distributions. Consequently, they play an important role in the description of stochastic processes. For example, the characteristic function of the Langevin dynamics in the y-direction, $Y(t) = \int_0^t \xi(t')dt'$, see Eq. (5.9), with $Y_0 = 0$ reads

$$\Phi(k,t) = \left\langle \exp\left[ik\int_0^t \xi(\tau)d\tau\right]\right\rangle. \tag{5.60}$$

The notion of the characteristic function can be generalized to a characteristic functional [97], $\Phi[f(\tau)] = \langle\exp[i\int f(\tau)\xi(\tau)d\tau]\rangle$. Using this expression, we can obtain all n-point correlation functions,

$$\langle\xi(t_1)\xi(t_2)\ldots\xi(t_n)\rangle = \frac{1}{i^n}\left[\frac{\delta^n\Phi[f(\tau)]}{\delta f(t_1)\delta f(t_2)\ldots\delta f(t_n)}\right]_{f=0}. \tag{5.61}$$

Equation (5.60) can be rewritten in the following form

$$\Phi(k,t) = \langle\exp[ikY(t)]\rangle = \int_{-\infty}^{\infty}\exp(iky)\langle\delta[y - Y(t)]\rangle dy. \tag{5.62}$$

Taking into account that $P_Y(y,t) = \langle\delta[y - Y(t)]\rangle$, we obtain that the pdf $P_Y(y,t)$ is, by definition, the inverse Fourier transform of the characteristic function,

$$\begin{aligned}
P_Y(y,t) &= \frac{1}{2\pi}\int_{-\infty}^{\infty}\Phi(k,t)\exp(-iky)dk \\
&= \frac{1}{2\pi}\int_{-\infty}^{\infty}\exp(-iky)\left\langle\exp\left[ik\int_0^t\xi_y(t')dt'\right]\right\rangle dk. \tag{5.63}
\end{aligned}$$

As a result, we obtain for the MSD

$$\langle\langle X^2(t)\rangle\rangle = \frac{D_x}{\pi}\int_0^t\left\langle\exp\left[ik\int_0^{t'}\xi_y(\tau)d\tau\right]\right\rangle dt'. \tag{5.64}$$

We will employ this expression to investigate more sophisticated dynamics with correlated noise.

5.3.1 *Gaussian colored noise: Correlation function*

Let the particle movement along the fingers be driven by a Gaussian colored noise $\xi_y(t)$ with arbitrary autocorrelation,

$$\langle\xi_y(t)\xi_y(t')\rangle = \gamma(t,t'). \tag{5.65}$$

White noise corresponds to the limiting case $\gamma(t, t') \to 2D_y \delta(t - t')$. The characteristic function of this zero-mean Gaussian random process is obtained via functional integration with the distribution

$$[d\rho(\xi_y)] = \exp\left[-\frac{1}{2}\int \xi_y(t') \Upsilon(t', t'') \xi_y(t'') dt\right][d\xi_y(t)]. \tag{5.66}$$

Here the functional differential is $[d\xi_y(t)] = \prod_\tau A(\tau) d\xi_y(\tau)$, $A(\tau)$ ensures the normalization condition $\int [d\rho(\xi_y)] = 1$, and $\Upsilon(t', t'')$ is related to correlation function, see below. To evaluate the Gaussian integral in Eq. (5.60) we consider its discretized form,

$$\left\langle \exp\left[ik\int_0^t \xi(\tau)d\tau\right]\right\rangle$$

$$\to I(\Upsilon, k) = A \int \prod_{j=1}^N d\xi_j \exp\left(-\frac{1}{2}\sum_{i,j}\xi_j \Upsilon_{i,j}\xi_j + \sum_j k_j\xi_j\right), \tag{5.67}$$

where $\xi_j = \xi(t_j)dt_j$, $\Upsilon_{i,j} = \Upsilon(t_i, t_j)$ is symmetric, $k_j = ik$, and $A = (2\pi)^{N/2} (\det \Upsilon)^{-1/2}$ is the normalization constant of the N dimensional Gaussian integral. The standard procedure of evaluating Gaussian integrals yields

$$I(\Upsilon, k) = \exp\left(\frac{1}{2}\sum_{i,j} k_i (\Upsilon^{-1})_{i,j} k_j\right). \tag{5.68}$$

Given that

$$\left.\frac{\partial^2 I(\Upsilon, k)}{\partial k_i \partial k_j}\right|_{k=0} = (\Upsilon^{-1})_{i,j} = \gamma(t_i, t_j), \tag{5.69}$$

and taking the continuous limit, we obtain

$$\Phi(k, t) = \left\langle \exp\left[ik\int_0^t \xi_y(\tau)d\tau\right]\right\rangle = \exp\left(-\frac{k^2}{2}\int_0^t\int_0^t \gamma(t', t'')dt'dt''\right). \tag{5.70}$$

The same expression can be obtained by considering the characteristic functional $\Phi[v(t)] = \exp\{iF[v(t)]\}$, where $F[v(t)]$ is the bilinear functional (5.22). It yields

$$\langle \xi_y(t)\xi_y(t')\rangle = \gamma(t, t')$$

$$= -\frac{\delta^2}{\delta v(t)\delta v(t')} \exp\left[\frac{i}{2}\int\int \gamma(\tau_1, \tau_2)v(\tau_1)v(\tau_2)d\tau_1 d\tau_2\right]\bigg|_{v=0}. \tag{5.71}$$

Taking $v(\tau) = k$, we obtain the characteristic function in Eq. (5.70).

5.3.2 Gaussian colored noise: MSD

We assume that the noise is stationary, i.e., $\gamma(t, t') = \gamma(t - t')$ and rewrite $\Phi(k, t)$ in the form

$$\Phi(k, t) = \exp\left[-k^2 \int_0^t \int_0^{t'} \gamma(t' - t'')dt''\,dt'\right] = \exp\left[-k^2 \int_0^t \gamma(t')(t - t')dt'\right].$$
(5.72)

The last exponential can be expressed by means of the Laplace transform $\mathcal{L}$ and the inverse Laplace transform $\mathcal{L}^{-1}$:

$$\int_0^t \gamma(t')(t - t')dt' \equiv \mathcal{L}^{-1}\mathcal{L}\int_0^t \gamma(t')(t - t')dt' = \mathcal{L}^{-1}\left[\frac{\hat{\gamma}(s)}{s^2}\right], \qquad (5.73)$$

where $\hat{\gamma}(s) = \mathcal{L}[\gamma(t)]$. Substituting the result into Eq. (5.64) for the MSD and performing the integration over k, we obtain

$$\langle\langle X^2(t)\rangle\rangle = \frac{D_x}{\sqrt{\pi}} \int_0^t \left\{\mathcal{L}_{t'}^{-1}\left[\frac{\hat{\gamma}(s)}{s^2}\right]\right\}^{-1/2} dt'. \qquad (5.74)$$

This expression represents a concise relation between the MSD of the overall ramified structure and the statistical characteristics of the stationary Gaussian noise, which drives the motion in the y-direction, in terms of the noise correlation function $\gamma(t)$.

In Ref. [100] the diffusion coefficient is defined as half the noise intensity,

$$2D_y = \int_0^\infty \gamma(t)dt = \hat{\gamma}(s = 0). \qquad (5.75)$$

If the integral in Eq. (5.75) is finite and nonzero, its value is the leading term in the expansion of $\hat{\gamma}(s)$ in the long time limit $s \to 0$. Up to the leading order we find that

$$\hat{\gamma}(s)/s^2 \simeq 2D_y/s^2 \quad \text{and} \quad \hat{\mathcal{L}}_{t'}^{-1}\left[\hat{\gamma}(s)/s^2\right] \simeq 2D_y t', \qquad (5.76)$$

and the asymptotic expression for the MSD in Eq. (5.74) reads

$$\langle\langle X^2(t)\rangle\rangle = \frac{2D_x}{\sqrt{2D_y\pi}}\sqrt{t} \quad \text{for } t \to \infty. \qquad (5.77)$$

In other words, the transport through the ramified structure is subdiffusive with the transport exponent $1/2$.

An interesting situation occurs when the Gaussian noise $\xi_y(t)$ has a power-law correlation function, for example, given by a Mittag–Leffler function,

$$\gamma(t) = A\lambda E_\alpha\left(-\lambda t^\alpha\right), \qquad (5.78)$$

where $0 < \alpha < 2$ and $A, \lambda > 0$ are arbitrary. In this case, Eq. (5.77) no longer holds and the MSD has to be determined from Eq. (5.74). Evaluating the inverse Laplace transform that appears in Eq. (5.74), we obtain

$$\mathcal{L}^{-1}\left[\frac{\hat{\gamma}(s)}{s^2}\right] = \mathcal{L}^{-1}\left[\frac{A\lambda s^{\alpha-3}}{s^\alpha + \lambda}\right] = A\lambda t^2 E_{\alpha,3}\left(-\lambda t^\alpha\right). \qquad (5.79)$$

In the long time limit this expression reduces to $4t^{2-\alpha}/\Gamma(3 - \alpha)$, and the calculation of the MSD yields

$$\langle\langle X^2(t)\rangle\rangle = \frac{2D_x}{\alpha}\sqrt{\frac{\Gamma(3 - \alpha)}{A\pi}}t^{\alpha/2}. \qquad (5.80)$$

A particular case occurs for $\alpha = 1$. Then the correlation function is exponential, $\gamma(t) = 4\lambda E_1(-\lambda t) = 4\lambda e^{-\lambda t}$, and the form of the MSD in Eq. (5.77) is exactly recovered on setting $\alpha = 1$ and $A = 4D_y$ in Eq. (5.80).

5.3.3 *Non-Gaussian white noise*

If particle motion along the fingers is driven by a non-Gaussian white noise $\xi_y(t)$, a so-called Lévy noise, then $Y(t)$, given in Eq. (5.9), is a generalized Wiener process [51, 2]. The random process $Y(t)$ has stationary independent increments on non-overlapping intervals [51], and it corresponds to the stable Lévy process with the characteristic function [51, 2, 3]

$$\Phi(k,t) = \exp\left(-D_y t |k|^\mu\right) \qquad (5.81)$$

with $0 < \mu < 2$. Substituting this result into Eq. (5.64) we obtain eventually the MSD

$$\langle\langle X^2(t)\rangle\rangle = \frac{2D_x}{\pi} \times \begin{cases} \dfrac{1}{D_y}\ln(t) & \text{if } \mu = 1, \\[2ex] \dfrac{\Gamma(1/\mu)t^{1-1/\mu}}{D_y^{1/\mu}(1 - 1/\mu)} & \text{if } \mu \neq 1. \end{cases} \qquad (5.82)$$

If $\mu = 2$, the MSD grows like $t^{1/2}$ as for a Gaussian noise. For $1 < \mu < 2$, the MSD displays subdiffusive behavior, and the transport exponent $1 - 1/\mu$ decreases as μ decreases from 2 to 1. When it reaches the value $\mu = 1$, the MSD grows ultraslowly. Finally, if $0 < \mu < 1$, the transport exponent is negative, and stochastic localization (transport failure) occurs [84].

Chapter 6

Inhomogeneous advection and superdiffusion

6.1 Introduction

In this chapter we present an example of a physical process that is analogous to a Lévy walk or a Lévy-like process. We consider transport that can be faster than ballistic, so-called hyper-diffusion, where the MSD is $\langle x^2(t) \rangle \sim t^\mu$, with a transport exponent $\mu > 2$. We discuss the behavior of observable quantities, e.g., the asymptotic form of the traveling packet, i.e., the tail of the distribution function. The physical mechanism of this "turbulent diffusion" is inhomogeneous convection, and we study the space–time evolution of the contaminant in the presence of inhomogeneous convective flow [24]. To emphasize the connection between inhomogeneous convection and hyper-diffusion, we consider transport on a subdiffusive substrate, e.g., on a comb structure. Our objective is to explain a possible mechanism of superdiffusion on the comb structure depicted in Fig. 1.1.

The convective flow is caused by external forces, and we consider the general case where the velocity of the convection flow is space dependent, i.e., the convection is inhomogeneous. For example, the anomalous diffusion might occur in the presence of external potentials [101] or in moving fluids [25, 27, 28]. We would like to determine how the observable shape of the initial packet changes, when the space-time evolution of the packet corresponds to convective flow. The fingers of the comb structure could be the result of partial sticking, which can occur when a tracer particle leaves the backbone and penetrates into ramified aquifers where fine channels exist alongside the main flow. This problem arises in a variety of applications, such as transport of pollutants in water flows through porous geological formations [102, 103], diffusion and reactions in porous catalysts [104], and fractal physiology [105, 106, 17]. It also relates to the more general formulation of Lévy walks [107, 108].

6.2 Superdiffusion on a comb structure

We determine the conditions for the inhomogeneous convection velocity such that the shape of the observable packet corresponds to normal diffusion, $\langle x^2(t) \rangle \sim t^\mu$ with $\mu = 1$, or to superdiffusion where $\mu > 1$. We assume that the inhomogeneity is given by a power-law function $\sim v_q x^q$. It turns out that Lévy-like processes take place if $q > 0$. The type of fractional dynamics caused by the inhomogeneous convective flow depends on the value of q. There exist three regimes: $0 < q < 1$, $q = 1$, and $q > 1$. If $0 < q < 1$, the transport exponent is arbitrary large, $1 < \mu < \infty$, corresponding to superdiffusion and hyper-diffusion, respectively. We call the process hyper-diffusion if the transport exponent is larger than 2, and the process is "beyond" the Lévy walk, i.e., a Lévy-like process. It is an important feature of this process that for $0 < q \leq 1$ all moments $\langle x^n(t) \rangle \sim t^{n\mu/2}$, $n = 1, 2, \dots$ are finite for finite times.

If $q > 1$, then all moments decay with time, and instead of diffusive spreading the transport results in localization due to "negative" inhomogeneous convection, negative in the sense that the convective flux is in the direction opposite to the contaminant spreading. The borderline case $q = 1$ corresponds to a very special type of spreading, where the distribution function of the particles is a log-normal distribution. The transport exponent approaches infinity, and the MSD is given by $\langle x^2(t) \rangle \propto t \exp(t)$.

6.2.1 *Diffusion vs subdiffusion*

We describe the evolution of the distribution function $P \equiv P(t, x, y)$ of a random walk on a comb via the Liouville equation or continuity equation,

$$\frac{\partial P(x, y, t)}{\partial t} + \operatorname{div} \mathbf{J} = 0, \tag{6.1}$$

where the current is given by

$$\mathbf{J} = \left(-D_x \delta(y) \frac{\partial P}{\partial x} + v_0 \delta(y), -D_y \frac{\partial P(x, y, t)}{\partial y} \right). \tag{6.2}$$

The current includes the usual diffusive flows and a term due to homogeneous advection along the backbone of the comb. Combs with homogeneous advection were first considered by Arkhincheev and Baskin [15] to model percolation in the presence of an electric field.[1] The corresponding advection-diffusion equation reads

$$\frac{\partial P(x,y,t)}{\partial t} - D_x \delta(y) \frac{\partial^2 P(x,y,t)}{\partial x^2} + v_0 \delta(y) \frac{\partial P(x,y,t)}{\partial x} - D_y \frac{\partial^2 P(x,y,t)}{\partial y^2} = 0$$

(6.3)

with the initial condition $P(x, y, t = 0) = \delta(x)\delta(y)$ and zero boundary conditions at infinity, $P(\pm\infty, y, t) = P(x, \pm\infty, t) = \partial_x P(\pm\infty, y, t) = \partial_y P(x, \pm\infty, t) = 0$.

Performing the Laplace and the Fourier transformations with respect to time and the x-coordinate, respectively, we obtain the solution (1.7) for subdiffusion with advection along the structure axis, i.e., the backbone,

$$\hat{\tilde{P}}(k, y, s) = \hat{\tilde{f}}(k, s)\hat{g}(y, s) = \exp\left(-\sqrt{s/D_y}|y|\right) \cdot \frac{1}{2\sqrt{sD_y} + ikv_0 + D_x k^2},$$

(6.4)

which can be written at $y = 0$ in terms of an auxiliary integration,

$$\hat{\tilde{P}}(k, y = 0, s) = \hat{\tilde{f}}(k, s) = \int_0^\infty \exp\left[-\left(2\sqrt{sD_y} + ikv_0 + D_x k^2\right)\tau\right] d\tau.$$

(6.5)

The inverse Fourier and Laplace transformations yield

$$P(x, y = 0, t) = f(x, t)$$
$$= \frac{D_y^{1/2}}{2\pi\sqrt{D_x t^3}} \int_0^\infty \tau^{1/2} \exp\left[-\frac{(x - v_0\tau)^2}{4D_x\tau} - \frac{D_y\tau^2}{t}\right] d\tau. \quad (6.6)$$

The total number of particles on the structure axis decreases with time,

$$\langle P(x, y = 0, t)\rangle = \int_{-\infty}^{\infty} P(x, 0, t) dx = \frac{1}{2\sqrt{\pi D_y t}}. \tag{6.7}$$

The solution (6.6) describes subdiffusion where the number of particles is not conserved. We will keep this point in mind when calculating the MSD.[2] This solution corresponds to transport along the structure axis with the MSD of the form

$$\langle x^2(t)\rangle = \frac{\int_{-\infty}^{\infty} x^2(t) P(x, 0, t) dx}{\langle P(x, 0, t)\rangle} = D_x \left(\frac{\pi t}{D_y}\right)^{1/2} + \frac{v_0^2}{D_y} t. \tag{6.8}$$

---

[1] In such a scenario, the Galilean invariance is broken during the sojourns of the particles inside the fingers. To some extent, it is the simplest model to introduce a Galilean variant fractional diffusion-advection equation [28].

[2] We can also consider the marginal distribution function $P_1(x, t) = \int_{-\infty}^{+\infty} P(x, y, t) dy$, which corresponds to the continuous time random walk model with conservation of the total number of particles. However, we prefer to consider $P(x, 0, t)$ here, since it is more appropriate for the purpose of this section.

In other words, for short times $\langle x^2(t) \rangle \sim D_x \left(\frac{\pi t}{D_y} \right)^{1/2}$ and the transport is subdiffusive. For large times, $\langle x^2(t) \rangle \sim \frac{v_0^2}{D_y} t$, and the transport corresponds to normal diffusion. However, the effective diffusion coefficient v_0^2/D_y is determined by the external forcing v_0. It is worth stressing that this is a nontrivial result, and it is reasonable to expect superdiffusion due to inhomogeneous convection of the form $v_q x^q$, with $q > 0$, in the x-direction.

6.2.2   *Superdiffusion as turbulent diffusion*

We consider now external forcing in the form of inhomogeneous convection along the backbone x-axis with the convection velocity[3] $v_x = v(x)\delta(y) = v_q x^q \delta(y)$. We will show later that the diffusive term in (6.3) can be neglected for simplicity. This simplification corresponds to a Liouville–Green asymptotics [110] or WKB approximation. In this section we provide some supporting evidence to this effect by considering the diffusion limit in Fourier-Laplace space.

In the case of inhomogeneous convection along the backbone axis, the current **J** reads

$$\mathbf{J} = (v_q x^q P(t,x,y)\delta(y) - D_x \delta(y)\partial P(t,x,y)/\partial x, -D_y \partial P(t,x,y)/\partial y) .$$
(6.9)

Substituting Eq. (6.9) into the Liouville equation (6.1), we find that the corresponding Fokker–Planck equation reads

$$\frac{\partial P}{\partial t} + \delta(y) v_q \frac{\partial x^q P}{\partial x} - D_x \delta(y) \frac{\partial^2 P}{\partial x^2} - D_y \frac{\partial^2 P}{\partial y^2} = 0,$$
(6.10)

where $q > 0$ and $P \equiv P(x,y,t)$.

To estimate the role of the diffusion term, it is convenient to work with dimensionless variables, which reduces the number of parameters in the equation to one. The dimensionless time and spatial coordinates are obtained by the re-scaling with relevant combinations of the comb parameters D_x and D_y. The dimensions of the parameters are $[D_x] = \text{cm}^3/\text{s}$ and $[D_y] = \text{cm}^2/\text{s}$. Consequently we obtain the following new variables for time, $D_y^3 t/D_x^2 \to t$, the spatial coordinates, $D_y x/D_x \to x$ and $D_y y/D_x \to y$, and the dimensionless velocity parameter $(D_x/D_y)^q v_q D_y^{-1} \to v_q$. The dimensionless form of Eq. (6.10) reads

$$\frac{\partial P}{\partial t} + \delta(y) v_q \frac{\partial x^q P}{\partial x} - \delta(y) \frac{\partial^2 P}{\partial x^2} - \frac{\partial^2 P}{\partial y^2} = 0.$$
(6.11)

---

[3]This can be caused by a change of the cross-sectional area of pipes or vessels in lymphatic or vascular systems, or by external potentials, see, e.g., [109, 101].

After the Laplace transformation with respect to time, we obtain the following solution,

$$\hat{P}(x, y, s) = f(x, s) \exp\left(-s^{1/2}|y|\right).$$ (6.12)

We have employed the equality,

$$\frac{\partial^2}{\partial y^2} \exp\left(-\sqrt{s}|y|\right) = [s - 2\sqrt{s}\delta(y)] \exp\left(-\sqrt{s}|y|\right).$$ (6.13)

Substituting Eq. (6.12) into Eq. (6.11), we obtain the following equation for $f \equiv f(x, s)$

$$- f'' + v_q x^q f' + q v_q x^{q-1} f + 2 s^{1/2} f = \delta(x),$$ (6.14)

with the boundary conditions, $f(\infty, s) = 0$ and $f(x, s) = 0$ for $x < 0$. Performing the Fourier transformation $\mathcal{F}[f(x, s)] = \tilde{f}(k, s)$, we obtain

$$k^2 \tilde{f}(k, s) + i k v_q \frac{\partial^q}{\partial k^q} S[\tilde{f}(k, s)] + 2\sqrt{s}\tilde{f}(k, s) = 1.$$ (6.15)

Here the fractional Riesz derivative is the result of the Fourier integration

$$\int_0^\infty (ix)^q f(x, s) e^{ikx} dx = \frac{\partial^q}{\partial k^q} \int_0^\infty f(x, s) e^{ikx} dx.$$ (6.16)

The functional $S[\tilde{f}(k, s)]$ is a short-hand notation for the following integral

$$S[\tilde{f}(k, s)] = \int_0^\infty \sin(kx + q\pi/2) f(x, s) dx.$$ (6.17)

The large scale asymptotics $x \gg 1$ correspond to $k \ll 1$ in Fourier space. Note that since x is dimensionless, the Fourier parameter k is dimensionless as well. The first term in Eq. (6.15) can be neglected under the condition

$$\lim_{\substack{k \to 0 \\ s \to 0}} \frac{k^2}{s^{1/2}} = 0.$$ (6.18)

This suggests that the diffusive term with D_x can indeed be neglected. [We provide a rigorous argument to this effect in Sec. 6.3.] This approximation depends on the form of the singularity of the convection velocity in the limit $x \to \infty$. In other words, the asymptotic solution of the homogeneous part of Eq. (6.15) for $k \ll 1$ depends on the exponent q in the power law x^q in Eq. (6.11) for $x \gg 1$ [111, 110]. In dimensioned coordinates, $x \gg D_x/D_y$. In this limit, the homogeneous part of Eq. (6.14) in dimensioned variables reads,

$$v_q x^q f' + q v_q x^{q-1} f + 2\left(D_y s\right)^{1/2} f = 0.$$ (6.19)

For $q < 1$, the solution of Eq. (6.19) has the following form,

$$f = \frac{1}{v_q x^q} \exp\left[-\frac{2(D_y s)^{1/2} x^{1-q}}{v_q(1-q)}\right],$$ (6.20)

where we have used that $g(x)\delta(x) = g(0)$.

For $q > 1$, the inverse Laplace transformation of Eq. (6.20) can be performed only if $v_q < 0$.[4] It is convenient to rewrite the solution in Laplace space, Eq. (6.20), in a symmetric form with respect to x:

$$f = \frac{1}{|v_q||x|^q} \exp\left[-\frac{2(D_y s)^{1/2}|x|^{1-q}}{|v_q(q-1)|}\right].$$ (6.21)

This solution is also valid for both $q < 1$ and $v_q > 0$. It corresponds to hyper-diffusion along the structure axis, where all moments exist and are determined by the following symmetrical distribution function for both $x > 0$ and $x < 0$,

$$P(x, y = 0, t) = \frac{D_y^{\frac{1}{2}}|x|^{1-2q}}{v_q^2(1-q)\sqrt{\pi t^3}} \exp\left[-\frac{D_y x^{2-2q}}{v_q^2(1-q)^2 t}\right].$$ (6.22)

From Eq. (6.22) we obtain the MSD, conditioned on a constant number of particles,

$$\frac{\langle x^2(t)\rangle}{\langle P(x, y = 0, t)\rangle} = \Gamma\left(\frac{2-q}{1-q}\right)\left[\frac{v_q^2(1-q)^2}{D_y}\right]^{\mu} t^{\mu},$$ (6.23)

where $\Gamma(z)$ is the gamma function. The transport exponent $\mu = 1/(1-q)$ corresponds to hyper-diffusion and it can be arbitrarily large, $1 < \mu < \infty$, for $q < 1$. Nevertheless, all moments are finite, as follows straightforwardly from Eq. (6.22). It is a kind of "turbulent diffusion," associated with accelerated Lévy walks. When $q = 0$, Eq. (6.22) corresponds to the result in Eq. (6.8). When $q > 1$, the situation is more interesting and leads to a new effect. In this case, the transport exponent is negative, and localization occurs. Recall that equations (6.20) and (6.22) were obtained by neglecting the diffusion term with D_x. Consequently, these solutions are not exact, but are relevant for the large scale asymptotics $x \gg D_x/D_y$. Asymptotic behavior for $q \geq 1$ requires special care, which will be dealt with in Sec. 6.3.

---

[4] For $q \geq 1$, the inverse Laplace transformation of Eq. (6.20) does not exist for $v_q > 0$.

6.2.3 *Exponential diffusion and log-normal distribution*

When $q = 1$, we deal with a borderline case that corresponds to an exponentially fast process as we will see below. Equation (6.14), which now reads in dimensioned form

$$v_1 x f' + [2(sD_y)^{1/2} + v_1]f = \delta(x), \qquad (6.24)$$

yields the so-called log-normal distribution. Since this equation is invariant with respect to the reflection $x \to -x$, it is valid for the whole x-axis from $-\infty$ to $+\infty$ with the same boundary conditions. The solution is given by

$$f(s, x) = \frac{\delta(x)}{2\sqrt{sD_y}} + \frac{C_1}{|x|^r}, \qquad (6.25)$$

where $r = a \equiv 1+2[sD_y]^{1/2}/v_1$, for $|x| > 1$ and $r = -a$ for $|x| \leq 1$. We used here that $-x\partial\delta(x)/\partial x = \delta(x)$ for a class of constant test functions. Note that the appearance of the $\delta(x)$ function in both Eq. (6.24) and its solution (6.25) exceeds the accuracy of the large scale asymptotics. However for the sake of mathematical rigor, we keep this function in both equations. This δ-function solution for the inhomogeneous equation can also be obtained by the Fourier transform. In this case, the specific solution in Fourier space is $f_k = (2\sqrt{sD_y})^{-1}$, and the inverse Fourier transformation yields the first term in (6.25). The constant C_1 is determined by the condition that the dimension of the second item in Eq. (6.25) agrees with the dimension of the first term. Consequently, $C_1 = (D/v_1)^r C_2$, where C_2 is given by the initial conditions after the inverse Laplace transformation.

To perform the inverse Laplace transformation, it is convenient to write the second term in Eq. (6.25) in the form of an exponential: $|z|^{-\sqrt{s}} = \exp(-\sqrt{s}\ln|z|)$. Then the inverse Laplace transformation becomes a standard procedure, and we obtain the following solution:

$$P(x, y = 0, t) = \frac{\delta(x)}{2\sqrt{\pi D_y t}} + \frac{D_y^{1/2}|\ln(|v_1 x/D_y|)|}{v_1^2|x|\sqrt{\pi t^3}} \exp\left[-\frac{D_y \ln^2(|v_1 x/D_y|)}{v_1^2 t}\right].$$
$$(6.26)$$

The first term corresponds to the smoothly time-decaying pinning distribution, while the second term, which is the most important for the asymptotic transport, corresponds to a kind of log-normal distribution. Again, we do not take into account the diffusive term, since an asymptotic space-time evolution of the initial distribution is determined by the drift component of the current, but not by diffusion.

We calculate the second moment in a manner similar to Eq. (6.8). Since $\langle P \rangle = 1/(2\sqrt{\pi D_y t})$, see Eq. (6.7), the MSD is

$$\langle x^2(t) \rangle = \frac{\langle x^2 P \rangle}{\langle P \rangle} \propto t \exp[v_1^2 t / D_y]. \tag{6.27}$$

This exponentially fast spreading results from the Lévy-like process, caused by the inhomogeneous displacement current in the system. It is straightforward to see from Eq. (6.26) that the flux (6.9) vanishes at infinity. The essential difference between the distribution for the subdiffusion described by Eq. (6.6), which is a stretched exponential, $P(|x|, y = 0, t) \propto \exp\left[-(x^4/t)^{1/3}\right]$ for $|x| \gg D_x/D_y$ and $t \gg D_y^3/D_x^2$, and the distribution for the superdiffusion described by Eq. (6.26) is obvious. Clearly, the main mechanism of superdiffusion on the comb structure is the inhomogeneous convection.

6.3 Liouville–Green approximation

In this Section we show that the approximation performed above, based on the condition (6.18), is adequate and corresponds to the Liouville–Green approximation, also called the WKB approximation.[5]

The total number of particles moving along the structure axis, the backbone, decreases with time due to the branch structure: $\langle P \rangle = \int_{-\infty}^{\infty} f(x,t)dx = 1/(2\sqrt{\pi D_y t})$. However, we can consider instead the CTRW for the total number of particles, which is described by the marginal pdf

$$P_1(x,t) = \int_{-\infty}^{\infty} P(x,y,t)dy, \tag{6.28}$$

and obtain the Liouville–Green asymptotic solution for the total number of particles, described by the reduced equation for $P_1(x,t)$ for $x > 0$. Substituting the marginal pdf into Eq. (6.13) we obtain

$$P_1(x,t) = \mathcal{L}^{-1}\left[\frac{1}{2\sqrt{D_y/s}} \hat{P}(x, y = 0, s) \right]. \tag{6.29}$$

---

[5]This approximation was used independently by Liouville and Green. In quantum mechanics this approximation is known as the Wentzel–Kramers–Brillouin (WKB) approximation. However, we quote [110]: "the contribution of these authors was not the construction of the approximation (which was already known), but the determination of connection formulas for linking exponential and oscillatory Liouville–Green approximations across a turning point on the real axis".

The solution (6.20) becomes the CTRW solution in Laplace space,

$$\hat{P}_1(x,s) = \frac{2}{v_q x^q s^{1/2}} \exp\left[-\frac{2s^{1/2}x^{1-q}}{|v_q(q-1)|}\right]. \qquad (6.30)$$

Taking this into account, we obtain the equation for P_1 by integrating Eq. (6.11) with respect to the variable y. Its homogeneous part (for $x \gg D_x/D_y$) reads in Laplace space, with $\hat{P}_1 \equiv \hat{P}_1(x,s)$:

$$s\hat{P}_1 + (D_y s)^{1/2}\left[v_q \frac{\partial x^q \hat{P}_1}{\partial x} - \frac{\partial^2 \hat{P}_1}{\partial x^2}\right] = s^{-1/2}\delta(x). \qquad (6.31)$$

Again, for convenience we consider this equation in dimensionless variables. This equation corresponds to the FFPE (1.20) with an inhomogeneous convection term and $\alpha = 1/2$. It is instructive to consider arbitrary $0 < \alpha < 2$, replacing $s^{1/2}$ with $s^{1-\alpha}$, and in dimensionless form we find

$$s\hat{P}_1 + s^{1-\alpha}\left[-\hat{P}_1'' + v_q x^q \hat{P}_1' + q v_q x^{q-1}\hat{P}_1\right] = 0. \qquad (6.32)$$

We remove the term with the first derivative by the substitution

$$\hat{P}_1 = s^{\alpha-1}\exp[v_q x^{q+1}/2(q+1)]w(x,s). \qquad (6.33)$$

We have $w'' = R(x)w$, where

$$R(x) = \frac{v_q^2 x^{2q}}{4}\left[1 + \frac{2q}{v_q}x^{-q-1} + \frac{4}{\alpha v_q^2}s^{\alpha}x^{-2q}\right]. \qquad (6.34)$$

The Liouville–Green approximation for w, which satisfies the boundary conditions for Eq. (6.11), is given by

$$w = BR^{-1/4}\exp\left[-\int R^{1/2}dx\right] = B\sqrt{\frac{v_q}{2}}\frac{1}{x^q}\exp\left[-\frac{v_q x^{q+1}}{2(q+1)} + \frac{s^{\alpha}x^{1-q}}{\alpha v_q(q-1)}\right], \qquad (6.35)$$

where B is a constant. Similarly, we obtain the Liouville–Green solution for negative $x \ll -1$. Therefore, taking $B = (2/v_q)^{3/2}$ and $\alpha = 1/2$, we find that Eq. (6.33) coincides exactly with the solution (6.30). This implies that removing the second derivatives from the Fokker–Planck equation, or the term k^2 in Fourier space in the limit $k \to 0$, corresponds to the Liouville–Green approximation for the Fokker–Planck equation with inhomogeneous (superdiffusive) convection.

6.3.1 *The Fox function solution*

Let us consider the solution (6.6) for the marginal pdf $\hat{P}_1(x, s)$ in Laplace space. With the help of Eq. (6.4), we can express the pdf in the form:

$$\hat{P}_1(x, s) = s^{-1/2}\mathcal{F}^{-1}\left[\frac{\left(\frac{D_x}{2\sqrt{D_y}}\right)^{-1}}{\left(k + \frac{iv_0}{2D_x}\right)^2 + \frac{v_0^2}{4D_x} + \left(\frac{D_x}{2\sqrt{D_y}}\right)^{-1}s^{1/2}}\right]. \tag{6.36}$$

For the inverse Fourier transformation we use the known Fourier transform of the Lorentzian,

$$\frac{1}{2\pi}\int_{-\infty}^{\infty} dk\, e^{ik(x+u)-r|x|} = \frac{1}{\pi}\frac{r}{(x+u)^2 + r^2}. \tag{6.37}$$

We can also define a generalized diffusion coefficient $D_{\frac{1}{2}} = D_x/(2\sqrt{D_y})$, see Eq. (1.20). Therefore, we obtain for the marginal pdf

$$P_1(x, s) = \frac{1}{2D_{\frac{1}{2}}}\exp\left(-\frac{v_0}{2D_x}x\right)s^{-3/4}h(z), \tag{6.38}$$

where $z = 1 + (v_0^2/4D_{\frac{1}{2}})s^{-1/2}$, and

$$h(z) = \frac{1}{\sqrt{z}}\exp[-\rho\sqrt{z}] \tag{6.39}$$

with $\rho = |x|\left(s^{1/2}/D_{\frac{1}{2}}\right)^{1/2}$. The function $h(z)$ can be represented in terms of the Fox H-function:

$$h(z) = \rho H_{1,0}^{0,1}\left[\rho z^{1/2}\,\middle|\,\begin{matrix}(-1, 1)\end{matrix}\right]. \tag{6.40}$$

The Taylor series expansion of $h(z)$ in a neighborhood of $z = 1$ yields the j-th derivative of $h(z)$ [44],

$$h^{(j)}(z) = \rho z^{-j}H_{1,1}^{1,2}\left[\rho z^{1/2}\,\middle|\,\begin{matrix}(0, 1/2)\\(-1, 1), (j, 1/2)\end{matrix}\right]. \tag{6.41}$$

For $z = 1$ we find,

$$h^{(j)}(z = 1) = \rho z^{-j}H_{1,1}^{1,2}\left[\rho\,\middle|\,\begin{matrix}(0, 1/2)\\(-1, 1), (j, 1/2)\end{matrix}\right]$$

$$= \frac{(-1)^j}{\sqrt{\pi}}H_{2,0}^{0,2}\left[\frac{\rho^2}{4}\,\middle|\,\begin{matrix}-\\(0, 1), (1/2 + j, 1)\end{matrix}\right]. \tag{6.42}$$

Thus, taking into account the Taylor series of $h(z)$ in a neighborhood of $z = 1$, we can express the marginal pdf in Laplace space in terms of the following series:

$$\hat{P}_1(x, s) = \frac{\exp\left(-\frac{v_0}{2D_x}x\right)}{\sqrt{4\pi D_{\frac{1}{2}}}}$$

$$\times \sum_{j=0}^{\infty} \left(\frac{v_0^2}{4D_{\frac{1}{2}}}\right)^j s^{-3/4-j/2} H_{2,0}^{0,2}\left[\frac{x^2}{4D_{\frac{1}{2}}}s^{1/2}\left|\begin{array}{c}-\\(0,1),(1/2+j,1)\end{array}\right.\right].$$

$$(6.43)$$

Using the definition of the Fox H-function in terms of the Mellin–Barnes integral, we obtain from the inverse Laplace transform of the marginal pdf [44],

$$P_1(x, t) = \frac{\exp\left(-\frac{v_0}{2D_x}x\right)}{\sqrt{4\pi D_{\frac{1}{2}}t^{1/2}}}$$

$$\times \sum_{j=0}^{\infty} \left(\frac{v_0^2}{4D_{\frac{1}{2}}}t^{1/2}\right)^j H_{2,0}^{1,2}\left[\frac{x^2}{4D_{\frac{1}{2}}t^{1/2}}\left|\begin{array}{c}(3/4+j/2,1/2)\\(0,1),(1/2+j,1)\end{array}\right.\right]. \quad (6.44)$$

If $v_0 = 0$, we arrive at the marginal pdf $P_1(x, t)$ for the standard comb model in Eq. (1.17) , see also Refs. [44, 45, 3],

$$P_1(x, t) = \frac{1}{\sqrt{4\pi D_{\frac{1}{2}}t^{1/2}}} H_{2,0}^{1,2}\left[\frac{x^2}{4D_{\frac{1}{2}}t^{1/2}}\left|\begin{array}{c}(3/4,1/2)\\(0,1),(1/2,1)\end{array}\right.\right]$$

$$= \frac{1}{2|x|} H_{1,0}^{1,1}\left[\frac{|x|}{\sqrt{D_{\frac{1}{2}}t^{1/2}}}\left|\begin{array}{c}(3/4,1/4)\\(0,1)\end{array}\right.\right]. \quad (6.45)$$

6.4   Classification of solutions

We have shown that the nature of the anomalous transport on a sub-diffusive substrate depends on the external forcing. We have considered the case of inhomogeneous convection, described by a power law, $J_x(x, t) = v_q x^q \delta(y) P(x, 0, t)$. Though we did not explicitly consider the case of negative q above, for $q < 0$ the solution (6.22) corresponds to sub-diffusion [112, 3], and the MSD (6.23) implies that the transport exponent is $\mu = 1/(1 + |q|) < 1$. If $0 < q < 1$, the nature of the transport is superdiffusive, see Eq. (6.23), where all moments are finite; this is a Lévy-like process. The case of homogeneous convection, i.e., $q = 0$, corresponds formally to normal diffusion as in Eq. (6.8), but the effective diffusion coefficient, v_0^2/D_y, is determined by the external forcing v_0. The borderline case, $q = 1$, displays two features. First, the distribution of the particles moving along the structure axis corresponds to the log–normal distribution, where one deals not with a sum of independent random variables but with their product instead [2]. As a result, exponentially fast transport occurs, see Eq. (6.27). Second, the solution satisfies the natural boundary conditions, where the flux of particles moving along the backbone equals zero at infinity for all times. If $q > 1$, the negative convection leads to stochastic localization.

There are many possible values for the parameters q and v_q of the convection flow $v(x)$ in Eq. (6.9). However, the Laplace transform narrows the range of parameters to $v_q > 0$, $q < 1$ and $v_q < 0$, $q > 1$, which result in physically acceptable realizations. These values are important for the continuity of the contaminant pdf. When $q > 1$, this continuity ensures that the inhomogeneous convection moves in the direction opposite to the natural spreading of the contaminant, which leads to stochastic localization. This effect is sketched in Fig. 6.1 where the natural spreading corresponds to an increasing width of the initial packet. Conversely, when $q < 1$, the continuity condition ensures that the inhomogeneous convection moves in the same direction as the contaminant. It should be stressed that this effect can be realized in a subdiffusive medium only.

Note that the relaxation (spreading) of the initial packet along the structure axis is caused by the diffusive term $D_x \delta(y) \partial P/\partial x$ in the current. Anomalous transport is due to the convective flow only on the asymptotically large scale $x \gg D_x/D_y$, where the diffusion process can be neglected. As shown above, this approximation corresponds to the Liouville–Green asymptotic solution (6.35).

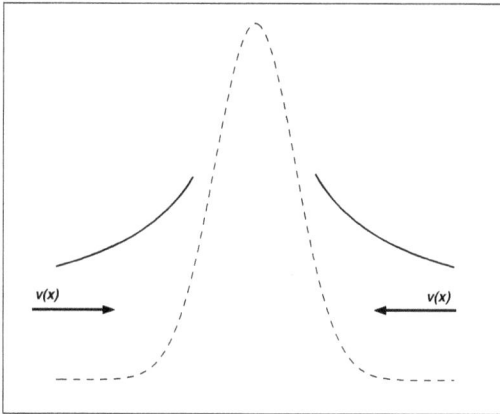

Fig. 6.1   A sketch of stochastic localization due to negative convection. The dashed curve corresponds to the spreading due to the relaxation process, while the solid lines are the tails of the power law due to superdiffusion. Reprinted figure with permission from A. Iomin and E. Baskin, Phys. Rev. E, 71, 061101 (2005). Copyright 2015 by the American Physical Society.

Chapter 7

Ultra-slow diffusion in 3D cylindrical comb

7.1 Introduction

We consider anomalous, ultra-slow diffusion in a three-dimensional cylindrical comb. This corresponds to the continuum limit of the situation shown in Fig. 7.1, i.e., a chain of circular discs of radius R, $0 < R \leq \infty$, threaded on the x axis. Ultra-slow phenomena have recently attracted considerable attention in biological search problems with long-range memories [113,114]. In CTRWs, ultra-slow diffusion occurs due to super-heavy-tailed distributions of the waiting times.[1] These heavy-tailed distributions result in the ratio $\langle x^2(t) \rangle / t^\alpha$ approaching zero as $t \to \infty$, in contrast with subdiffusion, where this ratio approaches a constant finite value. We show that the MSD for cylindrical combs shows logarithmic behavior in time, $\langle x^2(t) \rangle \sim \ln(t)$, which is a result of the transversal branch dynamics in the 2D space (see Chapter 5). This behavior has been discussed in the framework of scaling arguments for the return probability [119,34], which are based on the fractal dimension d_f of the transversal branch structure and the spectral dimension d_s. The latter is defined by the decay of the return probability $\sim t^{-d_s/2}$ [36,120], and $\langle x^2(t) \rangle \sim t^{1-\frac{d_s}{2}}$ for $d_s < 2$ [34]. As shown in Ref. [34], $\langle x^2(t) \rangle \sim \ln t$ for $d_s = 2$. Therefore, it is instructive to present a rigorous result by solving analytically the Fokker–Planck equation[2] in three-dimensional space within the cylindrical comb geometry constraint, where $d_f = 2$.

---

[1]For a detailed discussion, see Refs. [115–117], and for numerical results Refs. [117,118].
[2]We follow the macroscopic consideration of Ref. [121]. Further development of the theory in the framework of the mesoscopic random walk approach has been suggested in Ref. [122], see also Chapter 4.

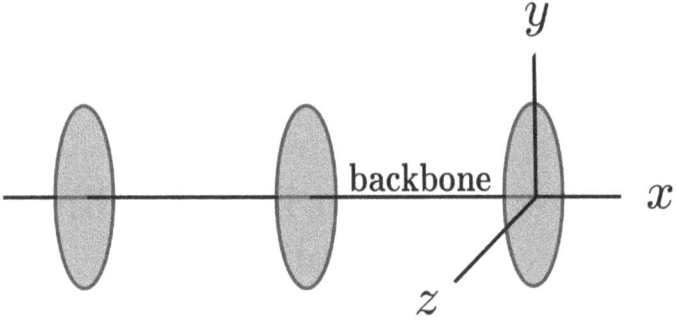

Fig. 7.1   Schematic representation of the 3D cylindrical comb. The backbone coincides with the x-axis and the discs, belonging to the y-z-plane, are threaded on the backbone and form a continuous cylinder. Reprinted from Chaos, Solitons & Fractals, 82, A. Iomin and V. Méndez, Does ultra-slow diffusion survive in a three dimensional cylindrical comb?, 142–147 (2016) with permission from Elsevier.

7.2   Dynamics in a cylindrical comb: Infinite discs

We consider a 3D cylindrical comb [123,34], shown in Fig. 7.1, in the framework of a standard formulation of the comb model (1.1) for the 3D case. The random dynamics on this structure is described by the 3D distribution function $P = P(x, y, z, t)$, where the x-axis corresponds to the backbone, while the dynamics on the two-dimensional secondary branches is described by the y and z coordinates. The non-dimenionalized diffusion equation reads

$$\frac{\partial P}{\partial t} = \delta(y)\delta(z)\frac{\partial^2 P}{\partial x^2} + D\left(\frac{\partial^2}{\partial y^2} + \frac{\partial^2}{\partial z^2}\right)P. \tag{7.1}$$

As shown in Chapter 6, it is convenient to work with dimensionless variables and parameters. These can be obtained by re-scaling the equation with relevant combinations of the comb parameters. We recall that $[D_x] = \mathrm{cm}^3/s$ and $[D_y] = [D_z] = \mathrm{cm}^2/s$. Therefore the dimensionless time and coordinates are

$$\frac{D_x^2}{D_y^3}t \to t, \quad \frac{D_x}{D_y}x \to x, \quad \frac{D_x}{D_y}y \to y/\sqrt{D}, \quad \frac{D_x}{D_z}z \to z/\sqrt{D}, \tag{7.2}$$

where D is introduced here as a dimensionless diffusion coefficient to specify the secondary branch dynamics. We first consider the case $R = \infty$ and assume natural boundary conditions at infinity, where the distribution function and its first spatial derivatives vanish. The initial conditions are given by

$$P_0 \equiv P(x, y, z, t = 0) = \delta(x)\delta(y)\delta(z). \tag{7.3}$$

It is sometimes convenient to consider the Fokker–Planck equation (7.1) in cylindrical coordinates:

$$\frac{\partial P}{\partial t} = \delta(y)\delta(z)\frac{\partial^2 P}{\partial x^2} + D\left[\frac{1}{r}\frac{\partial}{\partial r}\left(r\frac{\partial P}{\partial r}\right) + \frac{1}{r^2}\frac{\partial^2 P}{\partial \theta^2}\right], \qquad (7.4)$$

where r and θ are polar coordinates in the y-z-plane.

7.2.1   *Analysis in the time domain*

The formal solution of Eq. (7.1) can be presented as a convolution,

$$P(x, y, z, t) = \int_0^t G(y, z, t - t')F(x, t')dt', \qquad (7.5)$$

where $G(y, z, t)$ is the propagator for two-dimensional diffusion in the secondary branches, while $F(x, t)$ is related to the solution along the backbone.[3] Taking into account the cylindrical symmetry, one obtains for the y-z-plane

$$G(y, z, t) \equiv G(r, t) = \frac{1}{4\pi Dt}\exp\left(-\frac{r^2}{4Dt}\right), \qquad (7.6)$$

where $r^2 = y^2 + z^2$. To define the MSD in the x-direction, we need to obtain the marginal distribution $P_1(x, t)$ by integrating (7.5) with respect to y and z, taking into account that $dydz = d\theta r dr$. Equation (7.6) yields

$$
\begin{aligned}
P_1(x, t) &= \int_{-\infty}^{\infty} P(x, y, z, t)dydz \\
&= \int_0^t \left[\int_0^{\infty} rdr \int_0^{2\pi} d\theta G(r, t - t')\right] F(x, t')dt' \\
&= \int_0^t F(x, t')dt'.
\end{aligned}
\qquad (7.7)
$$

In Laplace space, this expression establishes a relation between $\hat{P}_1(x, s) = \mathcal{L}[P_1(x, t)]$ and $\hat{F}(x, s) = \mathcal{L}[F(x, t)]$, namely

$$\hat{F}(x, s) = s\hat{P}_1(x, s). \qquad (7.8)$$

[3]This relation in the time and Laplace domains is established in Eqs. (7.7) and (7.8). By analogy with the 2D comb, this formal presentation of the solution in convolution form follows from the Laplace transform of (7.1). It is tempting to understand this case in complete analogy with the 2D comb of Eq. (1.1), as well, where one assumes that the solution is the multiplication of the independent probability distribution functions in the Laplace domain $\hat{G}(y, z, s) \cdot \hat{F}(x, s)$. However, for the 3D comb the situation is more subtle. Namely, as we show in Sec. 7.2.2, the solution $\hat{G}(y, z, s)$ is singular at $(y, z) = (0, 0)$. Therefore, $\hat{F}(x, s)$ is not simply related to the distribution function $\hat{P}(x, y = 0, z = 0, s)$. However, in the time domain this relation can be easily established, using the formal solution in the convolution form (7.5).

The initial condition for the marginal distribution is $P_1(x, t = 0) = \delta(x)$. Using relation (7.8), we obtain an equation for $\hat{P}_1(x, t)$. Integrating Eq. (7.1) with respect to y and z, and taking into account Eqs. (7.5), (7.6) and (7.8), we find

$$s\hat{P}_1(x, s) = \frac{1}{4\pi D}\partial_x^2 \mathcal{L}[t^{-1}]s\hat{P}_1(x, s) + \delta(x). \tag{7.9}$$

Note that the Laplace transform of t^{-1} exists as a principal value integral [62]. We Fourier transform Eq. (7.9) and obtain

$$\hat{\tilde{P}}_1(k, s) = \frac{4\pi D}{s(4\pi D + k^2 \mathcal{L}[t^{-1}])}. \tag{7.10}$$

This yields the MSD in the form

$$\langle x^2(t)\rangle = \mathcal{L}^{-1}\left[-\frac{d^2}{dk^2}\hat{\tilde{P}}_1(k, s)\right]_{k=0} = \frac{1}{2\pi D}\int_{-i\infty}^{+i\infty}\mathcal{L}[t^{-1}]\frac{e^{st}ds}{s}. \tag{7.11}$$

Taking into account that $\mathcal{L}^{-1}\mathcal{L}\left[t^{-1}\right] \equiv t^{-1}$ and $e^{st}/s = \int^t e^{st}dt + C$, where C is an integration constant, we obtain

$$\langle x^2(t)\rangle = \frac{1}{2\pi D}\ln(t) + \frac{C}{2\pi D t} \sim \frac{1}{2\pi D}\ln(t) \text{ as } t \to \infty. \tag{7.12}$$

Therefore, the large time dynamics is governed by ultra-slow diffusion with the MSD growing like $\ln(t)$.

7.2.2   Consideration in the Laplace domain

We now study the relation between the temporal dynamics and the dynamics in Laplace space. Performing the Laplace transformation of Eq. (7.1), we obtain

$$s\hat{P} = \delta(y)\delta(z)\partial_x^2\hat{P} + D(\partial_y^2 + \partial_z^2)\hat{P} + P_0. \tag{7.13}$$

Correspondingly, in the Laplace domain Eq. (7.5) reads

$$\hat{P}(x, y, z, s) = \hat{G}(y, z, s)\hat{F}(x, s) \equiv \hat{G}(r, s)\hat{F}(x, s), \tag{7.14}$$

where we have taken into account the cylindrical symmetry in the last term. Using Eqs. (7.13) and (7.14), and introducing a new variable in the form of a scaled radius $u = r\sqrt{s/D}$, we find the solution for $\hat{G}(r, s) \equiv \hat{G}(u)$ from the equation

$$u^2\hat{G}'' + u\hat{G}' - u^2\hat{G} = 0, \tag{7.15}$$

where the prime denotes the derivative with respect to u. This is an equation for the modified Bessel functions $I_0(u)$ and $K_0(u)$, see, e.g., [124]. The solution that satisfies the boundary condition at infinity, $r = \infty$, is the modified Bessel function of the second kind,

$$\hat{G}(r, s) = AK_0\left(r\sqrt{s/D}\right).\tag{7.16}$$

Note that the inverse Laplace transform of $K_0\left(r\sqrt{s/D}\right)$ is exactly the solution $G(r, t)$ in Eq. (7.6),

$$\int_0^\infty \frac{1}{4\pi Dt} \exp\left(-\frac{r^2}{4Dt} - st\right) dt = \frac{1}{2\pi D}K_0\left(r\sqrt{s/D}\right).\tag{7.17}$$

Therefore, the normalization condition and the initial condition for $G(r, t)$ are satisfied with $A = 1/(2\pi D)$. Taking into account the solution (7.16), we establish the relation (7.8) by integrating Eq. (7.14) with respect to y and z. Using a property of integration of the modified Bessel function,

$$\int_0^\infty uK_0(au)du = 1/a^2,\tag{7.18}$$

we obtain

$$\hat{P}_1(x, s) = \hat{F}(x, s)\int dy dz\, \hat{G}(x, y, s) = \frac{\hat{F}}{D}\int_0^\infty K_0\left(r\sqrt{s/D}\right) rdr = \frac{\hat{F}(x, s)}{s}.$$
$$\tag{7.19}$$

This expression coincides exactly with the result in Eq. (7.8), where we also used the Laplace transform $\hat{P}_1(x, s) = 2\pi\hat{F}\int_0^\infty \hat{G}(r, s)rdr$ in Eq. (7.7).

We need address an important technical point of the analysis above: *namely Eq. (7.13) cannot be integrated with respect to y and z, because $\hat{G}(r, s)$ is singular at $r = 0$. In this case, $\hat{F}(x, s)$ cannot be defined by this procedure in the Laplace domain.[4] Therefore, to continue the analysis, one has to return to the time domain and repeat the analysis for the temporal dynamics, performed in Sec. 7.2.1. This situation differs crucially from the analysis for the 2D comb model in Eq. (1.1), where the finite expression for the MSD can be obtained in the Fourier–Laplace domain.*

7.3 Dynamics in a cylindrical comb with finite discs

To analyze diffusion on combs with finite discs, we consider reflecting boundary conditions at $r = R$, such that $\partial_r\hat{G}(r = R, s) = 0$. This implies that the probability current in the direction normal to the boundary

---

[4]It is worth noting that this does not mean that the solution does not exists, and that the singularity of $\hat{G}(r, s)$ at $r = 0$ does not imply the absence of the solution as well. Moreover, the existence of the Laplace transform (7.17) ensures the existence of the Laplace solution $\hat{G}(r, s)$ in the entire disc, including $r = 0$ as well.

surface vanishes. In this case, the solution of Eq. (7.15) is given by the modified Bessel function of the first kind $I_0(u)$. Namely, shifting the argument, $u \to u = (R-r)\sqrt{s/D}$, in Eq. (7.15), we obtain[5]

$$\hat{G}(r,s) = I_0\big[(R-r)\sqrt{s/D}\big]/I_0\big(R\sqrt{s/D}\big), \qquad (7.20)$$

which satisfies the boundary condition

$$\frac{d}{dr}I_0\big[(R-r)\sqrt{s/D}\big]\Big|_{r=R} = I_1\big[(R-r)\sqrt{s/D}\big]\Big|_{r=R} = 0, \qquad (7.21)$$

while for $r = 0$, we obtain $\hat{G}(0,s) = 1$. This corresponds to a standard construction of the solution [15,24]. In this case, integration of Eq. (7.13) with respect to y and z yields

$$s\hat{P}_1(x,s) = \partial_x^2 \hat{F}(x,s) + \delta(x). \qquad (7.22)$$

Again, the relation between $\hat{P}_1$ and $\hat{F}$ can be established by the integration of $I_0(u)$ over the y-z-surface of the discs,

$$2\pi \int_0^R rI_0\big[(R-r)\sqrt{s/D}\big]dr = 2\pi R^2 \int_0^1 uI_0[a(1-u)]du. \qquad (7.23)$$

We have used the following substitutions, $u = r/R$ and $a = R\sqrt{s/D}$. Using a further change of variables, $w = 1 - u$, we obtain two integrals [125],

$$\frac{R^2}{a}\int_0^a I_0(w)dw - R^2\int_0^1 wI_0(aw)dw = \frac{R^2}{a}\Big[2\sum_{n=0}^{\infty} I_{2n+1}(a) - I_1(a)\Big]. \qquad (7.24)$$

We are interested in the long time dynamics, when $s \to 0$ and $a \ll 1$, respectively. In this case, $I_n(a) \approx \left(\frac{a}{2}\right)^n/\Gamma(n+1)$. Consequently, we retain only the first term with $n = 0$ in the sum, which yields $I_1(a)$ in the square brackets. Finally, we obtain for the long time asymptotics[6]

$$2\pi \int_0^R \hat{G}(r,s)rdr \approx \pi R^2. \qquad (7.25)$$

This yields the following relation

$$\hat{F}(x,s) = \frac{\pi}{R^2}\hat{P}_1(x,s). \qquad (7.26)$$

---

[5]Obviously this corresponds to shifted arguments y and z in the Fokker-Planck equation (7.1) without touching the comb geometry, $\partial_t P = \delta(y)\delta(z)\partial_x^2 P + D[\partial_{(y_0-y)}^2 + \partial_{(z_0-z)}^2]P$, where $y_0^2 + z_0^2 = R^2$. Since y_0 and z_0 are constants, this shift does not change the dynamics, and this procedure is justified.
[6]This result can be obtained from Eq. (7.20) taking into account that if the argument is small, then $\hat{G}(r,s) \approx 1$ in the limit $s \to 0$.

Substituting relation (7.26) into Eq. (7.22) and performing the inverse Laplace transformation, we obtain the Fokker–Planck equation for normal diffusion with the diffusion coefficient π/R^2,

$$\frac{\partial P_1}{\partial t} = \frac{\pi}{R^2}\frac{\partial^2 P_1}{\partial x^2}. \tag{7.27}$$

It is worth stressing that this long time diffusion takes place only for times larger then a transient time $t > t_0$, where $t_0 = R^2/D$. This situation is different from the long-time asymptotics observed in Ref. [115].

This result is generic for combs with finite secondary branches (either fingers in the 2D comb, or discs in the 3D comb). However, the finite boundary conditions for the y-z-discs result in the destruction of ultra-slow diffusion in the x-direction, i.e., along the backbone. Mathematically, this fact follows immediately from the Laplace transform $\hat{G}(r, s)$, which depends on the boundary conditions. The ultra-slow motion takes place only for normal diffusion in infinite discs.

7.3.1 *Scenarios of transitions from ultra-slow to normal diffusion*

It is worth noting that the solution (7.20) is the easiest way to obtain the result on normal diffusion. However, it does not shed light on the transition from ultra-slow to normal diffusion. Further, it does not show that this ultra-slow motion is a transient phenomenon. To address these points, let us return to the initial Eq. (7.15),

$$u^2\hat{G}'' + u\hat{G}' - u^2\hat{G} = 0. \tag{7.28}$$

For the reflecting boundary conditions, the solution of Eq. (7.15) is found in the form of a superposition of the modified Bessel functions $I_0(u)$ and $K_0(u)$,

$$\hat{G}(u) = AK_0(u) + BI_0(u), \tag{7.29}$$

where the boundary condition yields

$$\hat{G}'(u)\Big|_{r=R} = -AK_1(a) + BI_1(a) = 0, \tag{7.30}$$

with $a = R\sqrt{s/D}$. This boundary condition establishes a relation between the coefficients $A = A(s)$ and $B = B(s)$. There are two possibilities for the coefficients; (*i*) $B = AK_1(a)/I_1(a)$ and A is a constant, and (*ii*) $A = BI_1(a)/K_1(a)$, and B is a constant. These yield two solutions with different combinations of the coefficients,

$$\hat{G}_1(u) = \frac{1}{2\pi D}K_0(u) + \frac{K_1(a)}{I_1(a)}I_0(u), \tag{7.31}$$

$$\hat{G}_2(u) = \frac{I_1(a)}{K_1(a)}K_0(u) + I_0(u). \tag{7.32}$$

However, this ambiguity between $\hat{G}_1(u)$ and $\hat{G}_2(u)$ is easily resolved by the physical meaning of the solutions on different time scales. The first solution $\hat{G}_1(u)$ is valid for initial times when $s \gg 1$. Since $R > r$, the second term of this solution can be neglected, and $\hat{G}_1(u) = K_0(u)/2\pi D$ corresponds to ultra-slow diffusion, studied in Sec. 2. This process is transient and takes place for $t \ll R^2/D$.

It is reasonable to consider the second solution for small arguments, i.e., $s \to 0$. In this case, the first term, of the order $s\ln(s)$, is much smaller than $I_0(u)$, which is of order 1. Integrating with respect to y and z, we find

$$2\pi \int_0^R rI_0(ar/R)dr = 2\pi R^2 I_1(a)/a. \tag{7.33}$$

In the limit $a \to 0$, the Bessel function is $I_1(a) \sim a/2$ [124]. Finally, we obtain exactly the result in Eq. (7.25), $\int_0^R G_2(r,s)d^2r = \pi R^2$ (here $d^2r \equiv dydz$). This solution corresponds to normal diffusion at times $t \gg R^2/D$, considered in the previous section.

These two solutions $\hat{G}_1(u)$ and $\hat{G}_2(u)$ correspond to the initial time scale and to the asymptotically large time scale, respectively. The corresponding solutions for the marginal distribution $P_1(x,t)$ in the form of ultra-slow diffusion (7.8) and normal diffusion (7.26) as the limiting cases, are anticipated as well. Both are obtained above in Eqs. (7.9) and (7.27), respectively.

Note that diffusion in the side branch discs can be anomalous as well. Does the ultra-slow diffusion survive in this case? In other words, does the logarithmic growth of the MSD still occur? The answer is it does not. We prove this statement in the next section.

7.4 Anomalous diffusion in discs

What happens to ultra-slow diffusion if diffusion in the transversal disks is anomalous and described by a memory kernel $\mathcal{K}(t)$? The comb model (7.1) now reads, written as an extension of the Generalized Master Equation, see Eq. (1.23),

$$\partial_t P(x,y,z,t) = \delta(y)\delta(z)\partial_x^2 P(x,y,z,t)$$
$$+ D(\partial_y^2 + \partial_z^2)\int_0^t \mathcal{K}(t-t')P(x,y,z,t')dt'. \tag{7.34}$$

The temporal kernel[7] $\mathcal{K}(t)$ is defined in the Laplace domain via a waiting time pdf $\psi(t)$ [3,83],

$$\hat{\mathcal{K}} = s\hat{\psi}(s)/[1 - \hat{\psi}(s)]. \tag{7.35}$$

Repeating the analysis in the Laplace domain of Sec. 7.2.2, we obtain the solution (7.16) in the form

$$\hat{G}(r, s) = AK_0(Br), \tag{7.36}$$

where $B = \sqrt{s/D\hat{\mathcal{K}}(s)}$ and A is a normalization constant. Integration of Eq. (7.18) yields

$$\hat{F}(x, s) = \frac{B^2}{A}\hat{P}_1(x, s). \tag{7.37}$$

As already discussed above in Sec. 7.2.2, $\hat{G}(r, s)$ is a singular function at $r = 0$. Therefore, as in Eq. (7.13), straightforward integration of Eq. (7.34) with respect to the y- and z-coordinates can be performed only in the real time domain. However, the function $G(r = 0, t)$ does exists and correspondingly $\mathcal{L}[G(r = 0, t)](s)$ exists as well, at least as a principal value integral like in Eq. (7.9). Therefore, to obtain the equation for P_1, we have to return to the time domain and integrate Eq. (7.34) with respect to y and z. To be specific, let us consider subdiffusion in the y-z-discs, described by $\hat{\psi}(s) = 1/[1 + (\tau s)^\alpha]$, which corresponds to the memory kernel

$$\hat{\mathcal{K}}(s) = s^{1-\alpha}/\tau^\alpha, \tag{7.38}$$

where τ is a dimensionless characteristic time scale and $0 < \alpha < 1$. In this case $B^2 = s^\alpha \tau^\alpha / D$. Therefore, in the limit $r \to 0$, we have $Br \ll 1$, and Eq. (7.36) reads for small arguments [124]

$$K_0[(bs)^{\alpha/2}] \approx \ln\frac{2}{\gamma} - \frac{\alpha}{2}\ln(bs), \tag{7.39}$$

where γ is Euler's constant and $b = \tau(r^2/D)^{1/\alpha}$. We perform the inverse Laplace transformation for small r,

$$G(r, t) = A\int_{-i\infty}^{i\infty} K_0[(bs)^{\alpha/2}]e^{st}ds \approx A\delta(t)\ln\frac{2}{\gamma} - Ab^{-1}\frac{\alpha}{2}\int_{-i\infty}^{i\infty}\ln(p)e^{pt_b}dp, \tag{7.40}$$

where $t_b = t/b$. The last term can be represented in a form of an integral, [62]

$$\int_{-i\infty}^{i\infty}\ln(p)e^{pt_b}dp = \frac{d}{dt_b}\int_{-i\infty}^{i\infty}p^{-1}\ln(p)e^{pt_b}dp = -\frac{1}{t_b}. \tag{7.41}$$

---

[7]Here we use notation $\hat{\mathcal{K}}(s)$ to distinguish the kernel from the modified Bessel functions. However $\hat{\mathcal{K}}(s) \equiv \hat{K}(s)$, which is defined in Chapter 1, see Eqs. (1.24) and (1.47).

Finally, we obtain for small r,

$$G(r,t) \approx \frac{A\alpha}{2} \cdot \frac{1}{t}. \qquad (7.42)$$

This result is independent of r and therefore, the limit $r = 0$ is correct. For this approximate solution, the constant $A = 1/(2\pi D)$ is taken to satisfy the limit $\alpha = 1$, which corresponds to solution (7.6) at $r = 0$.

Repeating the procedures of Sec. 7.2, namely performing first the integration with respect to y and z in Eq. (7.34) and then the Laplace transformation with respect to time, the Fourier transformation with respect to x, and taking into account the result of Eq. (7.37), we obtain a modification of Eq. (7.10). It reads for $\hat{\bar{P}}_1(k,s) = \mathcal{LF}[P_1(x,t)]$,

$$\hat{\bar{P}}_1(k,s) = \frac{1}{s + \bar{D}_\alpha s^\alpha k^2 \mathcal{L}[t^{-1}])}, \qquad (7.43)$$

where $\bar{D}_\alpha = \alpha \tau^\alpha/D$. Repeating the argument for deriving Eq. (7.12), we obtain for the MSD

$$\langle x^2(t) \rangle = \frac{2\bar{D}_\alpha}{\Gamma(1-\alpha)} \int_0^t \frac{\ln(t')dt'}{(t-t')^\alpha} \simeq D_\alpha t^{1-\alpha} \ln(t). \qquad (7.44)$$

We have neglected a term $\sim \gamma t^{1-\alpha}$, which is a slower contribution to anomalous diffusion than the term retained in Eq. (7.44). Here $D_\alpha = 2\bar{D}_\alpha/\Gamma(2-\alpha)$ is a generalized transport coefficient. Subdiffusion with the transport exponent $1 - \alpha$ is dominant. We conclude that ultra-slow diffusion $\sim \ln(t)$ takes place only for $\alpha = 1$, which can be realized as the result of normal diffusion in the infinite secondary branched discs.

7.5   Scenarios of the breakdown of ultra-slow diffusion

The exact analytical result on ultra-slow diffusion along the x-axis, i.e., the backbone, with the MSD of the order of $\ln(t)$, obtained by solving the Fokker–Planck equation in the 3D cylindrical comb geometry, demonstrates that the ultra-slow motion results from normal diffusion in the secondary discs of infinite radius. A technical point of the analysis is the singularity of $\hat{G}(r,s)$ at $r = 0$. The integration of the comb equations (7.1) and (7.34) with respect to the y- and z-coordinates is performed in the time domain, while the relation between the marginal pdf $P_1(x,t)$ and the backbone pdf $P(x,r=0,t)$ is established in the Laplace domain.

7.5.1 *Finite boundary conditions*

An important modification of the model is a choice of the boundary conditions at a finite radius of the discs, or finite boundary conditions, which is a realistic situation. In this case, the physical realization of the ultra-slow transport is restricted to the transient time scale $t < t_0 = R^2/D$. In the CTRW theory [3], this transient time is determined by the exponential decay of the tempered waiting time distributions.[8] In particular, the introduction of finite boundary conditions for the 2D comb model (1.1) is a geometrical realization of such tempered waiting time pdfs [128]. Before the time t_0, both the waiting times and subdiffusion correspond to the boundary condition at infinity, while for times $t > t_0$, the waiting times have a finite scale and normal diffusion takes place. The transition over time in this tempered waiting time distribution from the short time scale $t < t_0$ to the long time asymptotics $t > t_0$ is continuous [128]. For the 3D comb model, we found solutions (7.31) and (7.32), which contain both ultra-slow and normal diffusions: $\hat{G}(u) = A(s)K_0(u) + B(s)I_0(u)$. The reflecting boundary condition chooses the coefficients in such a way that for the short time scale $s \gg 1$ ($t < t_0$), the second term $B(s)I_0(u)$ vanishes. The first term $A(s)K_0(u)$ corresponds to the "super-heavy-tailed" waiting time pdf that eventually leads to ultra-slow diffusion along the backbone. The latter is a transient process, since it takes place only for $t < t_0$. In the opposite case of the large time asymptotics, the boundary condition leads to the vanishing of the first term $A(s)K_0(u)$ for $s \to 0$. This realizes the exponential time decay of the tempered waiting time pdf, which can be estimated by taking into account the first two terms in the expansion of $I_0(u)$ in Eq. (7.32) for $s \to 0$. In this case, $I_0\left(r\sqrt{s/D}\right) \simeq 1 + sr^2/4D$ [124]. Integrating over d^2r, we obtain relation (7.26) in the form

$$\hat{F}(x,s) = \frac{6D}{\pi R^3} \cdot \frac{\hat{P}_1(x,s)}{s + 12D/R^2}. \tag{7.45}$$

Laplace inversion yields the exponential decay of the tempered waiting time pdf,

$$\psi(t) = \frac{6D}{\pi R^3} \exp\left(-\frac{12Dt}{R^2}\right), \tag{7.46}$$

which is specified by the comb parameters[9] R and D and valid for large times $t \gg R^2/D$.

---

[8]This situation can take place in Lévy walks in random [107] and chaotic [126] systems; see also the recent discussion in Ref. [127].

[9]Note that R and D are dimensionless parameters, which are scaled by the transport properties D_x, D_y, and D_z of the medium.

7.5.2   *Subdiffusion in discs*

If the transport in the secondary branches is anomalous diffusion
(subdiffusion), anomalous transport becomes dominant along the backbone
as well. As a result, ultra-slow diffusion is replaced by anomalous transport
with the MSD $\sim t^{1-\alpha} \ln(t)$, which is more complex than the usual power
law subdiffusion, and we call it *enhanced subdiffusion*. This continuous
transition from the ultra-slow motion for $\alpha = 1$ to enhanced subdiffusion
with $0 < \alpha \leq 1$ is due to anomalous diffusion in the secondary branch
dynamics, which is controlled by the transport exponent α. Since the side-
branch y-z-dynamics acts as traps for the projected motion along the x-axis,
the analytical form of the solution $G(r, t)$ is crucial for the realization of
ultra-slow diffusion. The latter takes place only for the Gaussian solution
(7.6), when $\alpha = 1$. If $\alpha < 1$, the solution of $G(r, t)$ corresponds to subdiffusion in the form of the Fox function. This subdiffusion leads to less
extensive invasion of the side branches in comparison with normal diffusion,[10] which increases the probability to return to the backbone at $r = 0$.
Eventually, this enhances the transport along the backbone, and a continuous transition, as a function of α, from ultra-slow diffusion to enhanced
subdiffusion takes place. In the opposite case, when transport in the discs is
enhanced, for example by additional radial advection, one anticipates that
the transport along the backbone can be saturated. This result was obtained numerically for the 2D comb model in the framework of the CTRW
description [85]. From the CTRW point of view, the analytical expression
of the solution $G(r, t)$ relates to the form of the waiting time distribution for
the transport along the backbone. In particular, only the Gaussian solution
of $G(r, t)$ in Eq. (7.6) corresponds to the "super-heavy-tailed" waiting time
pdf, which leads to ultra-slow diffusion along the backbone. This mechanism can be helpful to model ecological processes like in the generalization
of an elephant random walk model [113].

---

[10]A detailed explanation of this phenomenon in the framework of first passage theory
can be found in Ref. [122].

PART 3

Reaction-Transport Processes in Combs

Chapter 8

Front propagation: Spiny dendrites

8.1 Comb models for transport along spiny dendrites

8.1.1 *Introduction*

Front propagation in reaction-transport equations, including fractional reaction-subdiffusion systems, is an active field of study as reflected in a variety of recent publications, see for example [16, 17, 129] and the recent monograph [83]. New experimental findings on calcium transport and reaction transport in neuroscience [130, 131] pose new questions as to the impact of the geometry on calcium transport and reactions in spiny dendrites and the extension of various reaction-transport models to the case of subdiffusion.

Comb-like models can mimic ramified structures such as spiny dendrites of neuron cells [16, 17, 129] and can be used to describe the movement and binding dynamics of particles inside the dendritic spines. Spiny dendrites are dendrites with small lateral protrusions, called dendritic spines, located on the surface of a neuron's dendrite. They can be found on the dendrites of most principal neurons in the brain, including the pyramidal neurons of the neocortex, the medium spiny neurons of the striatum, and the Purkinje cells of the cerebellum. In Figs. 8.1 and 8.2 we show pictures of a spiny dendrite. Dendritic spines contain postsynaptic densities and help transmit electrical signals to the neuron's cell body. They typically receive input from a single synapse of an axon. Most spines consist of a bulbous head of about 1μm and a thin neck of about 0.1μm, which connects the head of the spine to the shaft of the dendrite. The dendrites of a single neuron can contain hundreds or thousands of spines. Their heads have an active membrane, and as a consequence, they can sustain the propagation of an action potential with a rate that depends on the spatial density of spines [132].

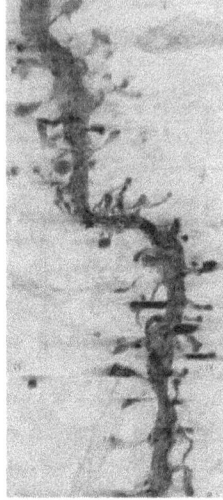

Fig. 8.1 Tomographic reconstruction of dendritic spines on a dendrite from a medium spiny neuron from the rat neostriatum. Brandner, D. and Withers, G., The Cell Image Library, CIL:39948. Available at `http://www.cellimagelibrary.org`. Accessed December 20, 2017/CC BY 3.0 `https://creativecommons.org/licenses/by/3.0/`

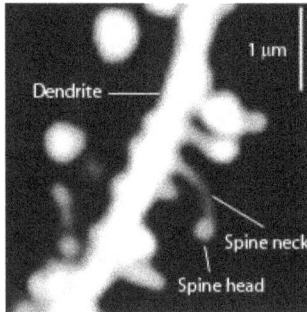

Fig. 8.2 Electron tomogram of a spiny dendrite. Wikimedia Commons `https://commons.wikimedia.org/wiki/File:Dendritic_spines.jpg`, Public Domain.

Decreased spine density can result in cognitive disorders, such as autism, mental retardation, and fragile X syndrome. Diffusion over branched, smooth dendritic trees is basically determined by classical diffusion, and the MSD along the dendritic axis grows linearly with time.

Recent experiments, together with numerical simulations, have investigated the calcium transport inside spiny dendrites to understand the role of calcium in signal transmission and neural plasticity.[1] Based on these experimental finding, different theoretical approaches have been developed to explore the transport properties of spiny dendrites. A seminal theoretical result on the front propagation of a solitary wave in dendritic spines was obtained in Ref. [135], where the spines are modeled as excitable clusters, while the dendrite is passive. An important finding is front propagation failure in terms of the separation between regularly spaced clusters. Propagation failure means that the overall velocity of the moving reaction front vanishes and that a static front solution exists. Many mechanisms can produce propagation failure, e.g., bistable kinetics, spatial heterogeneities, etc. [83].

The experimental observations of CaMKII (Ca^{2+}–calmodulin-dependent protein kinase II, a key regulator of synaptic function) translocation waves [130] raised important questions and led to various theoretical approaches to reaction front propagation. To study the translocation waves of CaMKII in spiny dendrites, a system of coupled reaction-diffusion equations was proposed in Ref. [131], where reactions between two chemical species were considered inside the dendrites and Markovian switching of particles between the dendrite and its spines [135, 131]. Propagation failure in terms of the switching rate was observed numerically [131].

For the analytical treatment of reaction-transport scenarios in spiny dendrites we explore both a linear reaction in spines and a nonlinear reaction along dendrites. The latter is a realization of the Fisher–Kolmogorov–Petrovskii–Piskunov (FKPP) scheme [136, 137]. As discussed in Refs. [130, 131], the process of activation is modeled by the reaction scheme $P_p + P_a \rightarrow 2P_a$, where P_p and P_a represent primed and activated CaMKII, respectively. Concentrations of the two species are denoted by the corresponding italic letters, i.e., P_p and P_a. This scheme is first order in P_p, first order in P_a, and second order overall. In terms of a FKPP interaction we deal with a term $P_p P_a$. Taking into account that the total CaMKII concentration is constant, we can explore the same scheme in the framework of a single nonlinear reaction-transport equation for the activated CaMKII, $P \equiv P_a$, with the kinetic rate term in the form of a logistic or FKPP-like rate term, $\bar{C}(P) = CP(1 - P)$, widely used in reaction-transport equations [83]. To account for the impact of the geometry on the reaction front propagation, we suggest a fractional subdiffusive comb model [16, 17], aug-

---

[1] This issue is reviewed in Refs. [133, 134], for example.

mented by the FKPP kinetic term. A hyperbolic scaling of the dynamical variables is applied to obtain a Hamilton–Jacobi equation and to estimate the overall velocity of the reaction front propagation.

8.1.2   Geometric impact on anomalous transport in dendrites

The spatial distribution of spines along the dendrite, their size and shapes are completely random [138], and not only the spine necks but the spine itself acts as a transport barrier. It is therefore reasonable to consider anomalous diffusion along both the spines and the dendrite [17]. Models are based on a comb-like structure that mimics a spiny dendrite, where the backbone is the dendrite and the fingers (lateral branches) are the spines, see Fig. 8.3, where we show both a smooth and a spiny dendrite. In this application, the dynamics inside the fingers corresponds to the spines, while the backbone describes diffusion along the dendrite.

Before discussing the CTRW description of spiny dendrites in the framework of a comb model, let us explain how anomalous diffusion in the comb model relates to the CaMKII transport along the spiny dendrite, and how the geometry of the latter relates to the anomalous transport. As discussed above, the spine cavities behave like traps for the kinase transport. It follows from a general treatment of Markov processes inside a finite region that the pdf of lifetimes inside a cavity with finite volume and arbitrary form decays exponentially with time t, see, for example, [77],

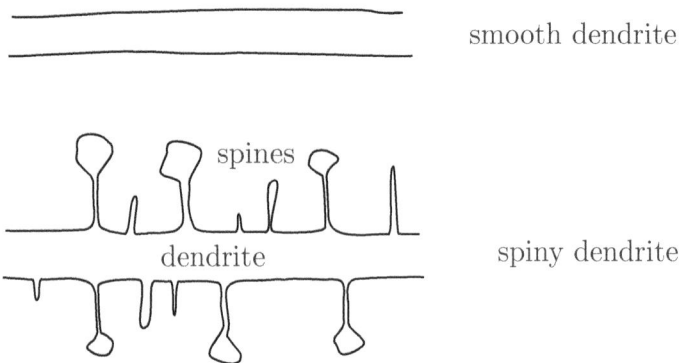

Fig. 8.3   Sketch of a smooth dendrite and a spiny dendrite. Reprinted from Chaos, Solitons & Fractals, 53, V. Méndez and A. Iomin, Comb-like models for transport along spiny dendrites, 46–51 (2013) with permission from Elsevier.

$\varphi(t) = \exp(-t/\tau)/\tau$. Here τ is a survival time (mean life time), defined by the minimum eigenvalue of the Laplace operator and determined by the geometry of the cavity. For example, as shown in Refs. [139,140], for spines with a head of volume V and a cylindrical spine neck of length L and radius a, the mean life time is $\tau = LV/\pi a^2 D = L^2/D$, where D is the diffusivity of the spine. Therefore, the mean probability to find a particle inside the spine after time t, i.e, the survival probability inside the cavity from 0 to t, averaged over all possible realizations of τ is given by the integral

$$\Psi(t) = \int_t^\infty \int_0^\infty \varphi(t'/\tau) f(\tau) d\tau dt', \tag{8.1}$$

where $f(\tau)$ is a distribution function of the survival times τ. Recall that the size and shape of spines are random [138]. Finally, the waiting time pdf can be easily calculated from Eq. (8.1), as follows,

$$\psi(t) = -\partial_t \Psi(t) = \int_0^\infty \varphi(t/\tau) f(\tau) d\tau. \tag{8.2}$$

In the simplest case, when the distribution is exponential $f(\tau) = (1/\tau_0) \exp(-\tau/\tau_0)$, we obtain from Eq. (8.2) that the general kinetics is non-Markovian and the waiting time pdf is a stretched exponential for large times,

$$\psi(t) = \frac{1}{\tau_0} \int_0^\infty \frac{e^{-t/\tau}}{\tau} e^{-\tau/\tau_0} d\tau = \frac{2}{\tau_0} K_0 \left(2\sqrt{t/\tau_0}\right)$$

$$\sim \left(\frac{t}{\tau_0}\right)^{-\frac{1}{4}} \exp(-\sqrt{t/\tau_0}), \quad t/\tau_0 \gg 1. \tag{8.3}$$

The situation is more interesting, when the distribution of the survival times is a power law, $f(\tau) \sim 1/\tau^{1+\alpha}$ with $0 < \alpha < 1$. In this case the waiting time pdf is a power law as well, $\psi(t) \sim 1/t^{1+\alpha}$, which leads to sub-diffusive motion along the dendrite. This result follows from CTRW theory, since all underlying micro-processes are independent Markovian ones with the same distributions, see, e.g., Ref. [3].

We now address the physical origin of a possible power law distribution for $\psi(t)$. We adapt some arguments from Ref. [36] to the present analysis. Consider the escape from a spine cavity from a potential point of view, where the geometrical parameters of the cavity can be related to a potential U. For example, for the simplest case, mentioned above Eq. (8.1), $U = VL/\pi a^2$ acts as potential barrier for the escaspe of a particle from the cavity, while D/τ_0 plays the role of the kinetic energy, or the "Boltzmann temperature". Therefore, the escape probability from the spine cavity or well is described by the Boltzmann factor $\exp(-U\tau_0/D)$. This value is proportional to the inverse waiting time, or survival time,

$$t \sim \exp\left(\frac{U\tau_0}{D}\right). \tag{8.4}$$

As explained above, the potential U is a random quantitiy distributed according to the Poisson distribution $P(U) = U_0^{-1}\exp(-U/U_0)$, where U_0 is an averaged geometrical spine characteristic. The probability to find the waiting time in the interval $(t, t+dt)$ is equal to the probability to find the trapping potential in the interval $(U, U+dU)$, namely $\psi(t)dt = P(U)dU$. We obtain from Eq. (8.4),

$$\psi(t) \sim \frac{1}{t^{1+\gamma}}. \tag{8.5}$$

Here $\gamma = D/(U_0\tau_0) \in (0,1)$ establishes a relation between the geometry of the dendrite spines and the subdiffusion observed in Refs. [141, 142]. It supports the application of the comb model, which is a convenient tool to explore analytically the anomalous transport in spiny dendrites in the framework of the CTRW approach.

8.1.3   *Anomalous diffusion in spines: I*

We apply the comb model to anomalous diffusion in spiny dendrites, which implies that the flux of particles along the dendrite is given by

$$J_x = -D_x\delta(y)\frac{\partial P}{\partial x}. \tag{8.6}$$

The transport along the spines is anomalous due to the trapping process that occurs inside the spine, and the flux along the fingers can be written as,

$$J_y = -D_y\mathcal{D}^{1-\gamma}\frac{\partial P}{\partial y}, \tag{8.7}$$

where $P(x, y, t)$ is the density of particles, and $\mathcal{D}^{1-\gamma} = D D^{-\gamma}$ is a shorthand notation for the Riemann–Liouville fractional derivative (2.98a), see Chapter 2. It is convenient to define the latter by means of the Laplace transform (2.110),

$$\mathcal{L}\left[D^{-\gamma}f(t)\right] = s^{-\gamma}\hat{f}(s). \tag{8.8}$$

Inside a spine, the transport process is anomalous with the MSD $\langle y^2(t) \rangle \sim t^\gamma$, where $\gamma \in (0,1)$. Making use of the continuity equation for the total number of particles,

$$\frac{\partial P}{\partial t} + \mathrm{div}\mathbf{J} = 0, \tag{8.9}$$

where $\mathbf{J} = (J_x, J_y)$, we obtain the following evolution equation for transport along the spiny dendrite,

$$\frac{\partial P}{\partial t} - D_x \delta(y) \frac{\partial^2 P}{\partial x^2} - D_y \mathcal{D}^{1-\gamma} \frac{\partial^2 P}{\partial y^2} = 0. \tag{8.10}$$

The Riemann–Liouville fractional derivative in Eq. (8.10) is not convenient for carrying out the Laplace transformation of the equation. To ensure feasibility of the Laplace transformation, the preeminent tool for dealing with fractional equations, we reformulate the problem in a more suitable form.

8.1.4   Auxiliary: 3D comb

To shed light on this situation, we consider a comb in 3D [36]. This model is described by the distribution function $P_3 = P_3(x, y, z, t)$, whose evolution is given by the equation[2]

$$\frac{\partial P_3}{\partial t} - D_x \delta(y)\delta(z) \frac{\partial^2 P_3}{\partial x^2} - D_y \delta(z) \frac{\partial^2 P_3}{\partial y^2} - \frac{\partial^2 P_3}{\partial z^2} = 0. \tag{8.11}$$

It should be stressed that the z coordinate, not to be confused with the z variable introduced in the previous chapter for the cylindrical comb, is an auxiliary, virtual variable, introduced to describe fractional motion in the spines by means of a Markovian process. The actual distribution of interest is

$$P(x, y, t) = \int_{-\infty}^{\infty} P_3(x, y, z, t) dz, \tag{8.12}$$

which obeys the evolution equation,

$$\frac{\partial P}{\partial t} - D_x \delta(y) \frac{\partial^2 P_3(z=0)}{\partial x^2} - D_y \frac{\partial^2 P_3(z=0)}{\partial y^2} = 0. \tag{8.13}$$

The relation between $P(x, y, t)$ and $P_3(x, y, z = 0, t)$ can be expressed through their Laplace transforms,

$$\hat{P}_3(x, y, z = 0, s) = \frac{\sqrt{s}}{2} \hat{P}(x, y, s), \tag{8.14}$$

where $\hat{P}(x, y, s) = \mathcal{L}[P(x, y, t)]$ and $\hat{P}_3(x, y, z, s) = \mathcal{L}[P_3(x, y, z, t)]$. If we apply the Fourier-Laplace transformation to Eq. (8.12), we obtain

$$\hat{\tilde{P}}(k_x, k_y, s) = \hat{\tilde{P}}_3(k_x, k_y, k_z = 0, s). \tag{8.15}$$

---

[2]Note that this geometry differs completely from ultra-slow diffusion in cylindrical comb, considered in Chapter 7.

We see that Eq. (8.14) represents a relation between $\hat{\hat{P}}_3\left(k_x, k_y, k_z = 0, s\right)$ and $\hat{\hat{P}}_3\left(k_x, k_y, z = 0, s\right)$. To find $\hat{\hat{P}}_3\left(k_x, k_y, k_z, s\right)$, we apply the Fourier–Laplace transformation to Eq. (8.11). After collecting terms, we find

$$\hat{\hat{P}}_3\left(k_x, k_y, k_z, s\right)$$
$$= \frac{1 - D_x k_x^2 \hat{\hat{P}}_3\left(k_x, y = 0, z = 0, s\right) - D_y k_y^2 \hat{\hat{P}}_3\left(k_x, k_y, z = 0, s\right)}{s + k_z^2}. \quad (8.16)$$

We have assumed, for simplicity, that the initial condition is given by $P_3\left(x, y, z, t = 0\right) = \delta\left(x\right)\delta\left(y\right)\delta\left(z\right)$. Setting $k_z = 0$, we obtain

$$\hat{\hat{P}}_3\left(k_x, k_y, k_z = 0, s\right)$$
$$= \frac{1 - D_x k_x^2 \hat{\hat{P}}_3\left(k_x, y = 0, z = 0, s\right) - D_y k_y^2 \hat{\hat{P}}_3\left(k_x, k_y, z = 0, s\right)}{s}. \quad (8.17)$$

Applying the inverse Fourier transformation to Eq. (8.16) we find

$$\hat{\hat{P}}_3\left(k_x, k_y, z, s\right)$$
$$= \frac{1 - D_x k_x^2 \hat{\hat{P}}_3\left(k_x, y = 0, z = 0, s\right) - D_y k_y^2 \hat{\hat{P}}_3\left(k_x, k_y, z = 0, s\right)}{2\sqrt{s}} e^{-\sqrt{s}|z|}. \quad (8.18)$$

Setting $z = 0$ yields

$$\hat{\hat{P}}_3\left(k_x, k_y, z = 0, s\right)$$
$$= \frac{1 - D_x k_x^2 \hat{\hat{P}}_3\left(k_x, y = 0, z = 0, s\right) - D_y k_y^2 \hat{\hat{P}}_3\left(k_x, k_y, z = 0, s\right)}{2\sqrt{s}}. \quad (8.19)$$

Combining Eqs. (8.17) and (8.19), we find

$$\hat{\hat{P}}_3\left(k_x, k_y, z = 0, s\right) = \frac{\sqrt{s}}{2} \hat{\hat{P}}_3\left(k_x, k_y, k_z = 0, s\right). \quad (8.20)$$

The inverse Fourier transformation with respect to k_x and k_y yields Eq. (8.14). Finally, performing the Laplace transformation of Eq. (8.13), we obtain

$$s\hat{P}(x, y, s) - D_x \delta(y)\frac{\partial^2 \hat{P}_3(x, y, z = 0, s)}{\partial x^2} - D_y \frac{\partial^2 \hat{P}_3(x, y, z = 0, s)}{\partial y^2}$$
$$= P(x, y, t = 0). \quad (8.21)$$

Employing the relation (8.14), dividing by $\sqrt{s}$, and then performing the inverse Laplace transformation, we arrive at the comb model transport equation,

$$\frac{\partial^{1/2} P}{\partial t^{1/2}} - D_x \delta(y)\frac{\partial^2 P}{\partial x^2} - D_y \frac{\partial^2 P}{\partial y^2} = 0. \quad (8.22)$$

Here $2D_{x,(y)} \to D_{x,(y)}$, and we used the short-hand notation for the Caputo fractional derivative, $\partial^\gamma / \partial t^\gamma$, (2.98b), which can be defined by the Laplace transform for $\gamma \in (0,1)$, see Sec. 2.6,

$$\hat{\mathcal{L}}\left[\frac{\partial^\gamma f}{\partial t^\gamma}\right] = s^\gamma \hat{f}(s) - s^{\gamma-1} f(t=0). \tag{8.23}$$

8.1.5 *Anomalous diffusion in spines: II*

Equation (8.22) describes a system where fractional transport takes place in both the dendrite x-direction and the spines y-direction. To render the diffusion in the dendrite normal, we apply the fractional integration $I_t^{1-\gamma} \equiv \mathcal{D}^{\gamma-1}$, defined by means of the Laplace transform (8.8), $\mathcal{L}\left[I_t^{1-\gamma} f(t)\right] = s^{\gamma-1}\hat{f}(s)$, to the diffusion term in the x-direction. After generalizing $1/2 \to \gamma \in (0,1)$, we obtain

$$\frac{\partial^\gamma P}{\partial t^\gamma} - D_x \delta(y) I_t^{1-\gamma} \frac{\partial^2 P}{\partial x^2} - D_y \frac{\partial^2 P}{\partial y^2} = 0. \tag{8.24}$$

Performing the Fourier–Laplace transformation of Eq. (8.24), we find

$$\hat{\tilde{P}}(k_x, k_y, s) = \frac{\hat{\tilde{P}}(k_x, k_y, t=0) - D_x k_x^2 \hat{\tilde{P}}(k_x, y=0, s)}{s + D_y k_y^2 s^{1-\gamma}}, \tag{8.25}$$

where the Fourier–Laplace transform of the distribution function is defined by its arguments $\mathcal{L}\mathcal{F}_x\mathcal{F}_y[P(x,y,t)](k_x, k_y, s) = \hat{\tilde{P}}(k_x, k_y, s)$. If $P(x,y,t=0) = \delta(x)\delta(y)$, the inverse Fourier transformation with respect to k_y yields

$$\hat{\tilde{P}}(k_x, y, s) = \frac{1 - D_x k_x^2 \hat{\tilde{P}}(k_x, y=0, s)}{s^{(2-\gamma)/2}\sqrt{D_y}} \exp\left(-|y| s^{\gamma/2}/\sqrt{D_y}\right). \tag{8.26}$$

Setting $y=0$ in Eq. (8.26) provides

$$\hat{\tilde{f}}(k_x, s) \equiv \hat{\tilde{P}}(k_x, y=0, s) = \frac{1}{s^{(2-\gamma)/2}\sqrt{D_y} + D_x k_x^2}, \tag{8.27}$$

which yields, after substituting it into Eq. (8.25),

$$\hat{\tilde{P}}(k_x, k_y, s) = \frac{1}{s + D_y k_y^2 s^{1-\gamma}}\left(1 - \frac{D_x k_x^2}{s^{(2-\gamma)/2}\sqrt{D_y} + D_x k_x^2}\right). \tag{8.28}$$

We can calculate the density of particles at a given point x of the dendrite at time t, namely $P(x,t)$, by integrating with respect to the y-coordinate in Fourier space,

$$\hat{\tilde{f}} = \hat{\tilde{P}}(k_x, s) = \hat{\tilde{P}}(k_x, k_y=0, s) = \frac{s^{-\gamma/2}\sqrt{D_y}}{s^{(2-\gamma)/2}\sqrt{D_y} + D_x k_x^2}. \tag{8.29}$$

Then

$$\langle x^2(s) \rangle = -\left.\frac{\partial^2}{\partial k_x^2}\hat{\bar{f}}(k_x, s)\right|_{k_x=0} = \frac{2D_x}{\sqrt{D_y}}\frac{1}{s^{2-\gamma/2}}, \tag{8.30}$$

and

$$\langle x^2(t) \rangle = \frac{2D_x}{\sqrt{D_y}}t^{1-\gamma/2}. \tag{8.31}$$

Equation (8.31) predicts subdiffusion along the spiny dendrite, which is in agreement with the experimental results reported in Ref. [141]. Note that this result is counterintuitive. Indeed, subdiffusion in spines, or fingers, should lead to slower subdiffusion in dendrites, or backbone, with the transport exponent smaller than in usual combs, since usually these two processes are strongly correlated. However, this correlation is broken due to the fractional integration $I_t^{1-\gamma}$ in Eq. (8.24), which turns diffusion along the backbone into normal diffusion. In other words, the different nature of the transport processes in the dendrite and the spines breaks the strong correlation. On the other hand, if we perform the inverse Fourier–Laplace transformation of (8.29), we obtain the fractional diffusion equation for $f(x,t)$:

$$\frac{\partial^{1-\gamma/2}f}{\partial t^{1-\gamma/2}} = \frac{D_x}{\sqrt{D_y}}\frac{\partial^2 f}{\partial x^2}. \tag{8.32}$$

This equation is equivalent to the Generalized Master Equation (1.23) in the diffusion limit,

$$\frac{\partial f}{\partial t} = \int_0^t K(t-t')\frac{\partial^2 f(x,t')}{\partial x^2}dt', \tag{8.33}$$

if the Laplace transform of the memory kernel is given by

$$\hat{K}(s) = (D_x/\sqrt{D_y})s^{\gamma/2}. \tag{8.34}$$

The latter corresponds to the waiting time pdf in Laplace space given by

$$\hat{\psi}(s) = \frac{1}{1 + (\sqrt{D_y}/D_x)s^{1-\gamma/2}}, \tag{8.35}$$

which behaves like $\psi(t) \sim t^{-2+\gamma/2}$ as $t \to \infty$. Let us employ the notation for a dynamical exponent d_w used in Refs. [141, 142]. If $d_w = 4/(2-\gamma)$, then the MSD grows like t^{2/d_w}. On the other hand, it has been found in experiments that d_w increases with the density of spines, ρ_s, and simulations show that d_w grows linearly with ρ_s. The experimental data are compatible with almost any dependence where d_w increases with ρ_s, due to the high variance of the data, see Ref. [142]. Equation (8.31) also establishes a phenomenological relation between the second moment and ρ_s. If the density spines is zero, then $\gamma = 0$ and $d_w = 2$ and normal diffusion occurs. If the spine density ρ_s increases, the anomalous exponent of the pdf (8.35), $1 - \gamma/2 = 2/d_w$, must decrease, i.e., the transport becomes more subdiffusive due to the increase of ρ_s, so that d_w has to increase as well. Our model predicts qualitatively that d_w increases with ρ_s, in agreement with the experimental results in Ref. [142].

8.2 Fractional reaction-diffusion along spiny dendrites

Geometrically, spiny dendrites in 3D space are completely described by a comb structure in 2D, where the spine density on the cylindrical surface of the dendrite is projected onto a 1D-axis, say the x-axis: $\rho(x, r = const, \theta) \to \rho(x)$. Here, $\rho(x, r = const, \theta)$ is the spine density, while $\rho(x)$ is the density of the comb fingers. In what follows, we consider $\rho(x) = g = const$, which appears to be the most realistic case. Fractional diffusion inside the spines is described by fractional diffusion inside the fingers. We consider a two-sided comb model as in Fig. 1.1, and the mathematical starting point of the phenomenological approach is the Fokker–Planck equation obtained in [16]:

$$\frac{\partial P}{\partial t} - \delta(y)\frac{\partial^2 P}{\partial x^2} - g\frac{\partial^2 P}{\partial y^2} = 0. \tag{8.36}$$

This equation is obtained by a re-scaling with relevant combinations of the comb parameters D_x and D_y, such that the dimensionless time and coordinates are $D_x^3 t / D_y^2 \to t$ $D_x/D_y \to x$, and $D_x/D_y \to y/\sqrt{g}$, respectively, see Chapter 6. The parameter g can be understood as a constant density of the fingers.

As discussed above, a variety of interactions inside the spines leads to correlated noises in dendritic spines [143]. The strong correlations lead to anomalous (subdiffusive) motion inside the spines. In a phenomenological description based on CTRWs, this subdiffusion is controlled by a waiting-time pdf, $\psi(t)$, decaying according to a power law. Normal diffusion in the spines of the particle density $P(x, y, t)$, for example activated CaMKII, is replaced by the anomalous transport term, see Eq. (1.23),

$$g\frac{\partial^2 P}{\partial y^2} \to g \int_0^t K(t - t')\frac{\partial^2 P(t')}{\partial y^2}dt'. \tag{8.37}$$

Here $K(t)$ is again the time memory kernel of the Generalized Master Equation and is defined in Eq. (8.34). Subdiffusion, where $\psi(t) = 1/(1 + t^{1+\gamma})$ with $0 < \gamma < 1$, yields $\hat{K}(s) = s^{1-\gamma}$ [144].

Equation (8.37) is a formal expression for the anomalous transport with a rather complicated form in the time domain, which is not convenient for analytical treatment. It is advantageous to investigate the comb model in Laplace space. Substituting Eq. (8.37) into Eq. (8.36), then performing the Laplace transformation and taking into account Eq. (8.34), we obtain the comb model in the Laplace domain,

$$s\hat{P} = \delta(y)\frac{\partial^2 \hat{P}}{\partial x^2} - gs^{1-\gamma}\frac{\partial^2 \hat{P}}{\partial y^2} + P_0. \tag{8.38}$$

Here $P_0 = P(x, y, t = 0)$ is the initial condition. As discussed above, the kernel $\hat{K}(s)$ is problematic for the inverse Laplace transformation, since it leads to the appearance of the initial condition. To overcome this obstacle, we multiply Eq. (8.38) by $s^{\gamma-1}$ and then perform the inverse Laplace transformation which yields,

$$\int_0^t (t - t')^{-\gamma} \left[\frac{\partial P(x, y, t')}{\partial t'} - \delta(y)\frac{\partial^2 P(x, y, t')}{\partial x^2} \right] dt' = g\frac{\partial^2 P(x, y, t)}{\partial y^2}. \tag{8.39}$$

Adding a reaction term to this equation, we obtain the reaction-transport equation:

$$\int_0^t (t - t')^{-\gamma} \left[\frac{\partial P(x, y, t')}{\partial t'} - \delta(y)\frac{\partial^2 P(x, y, t')}{\partial x^2} \right] dt'$$
$$= g\frac{\partial^2 P(x, y, t)}{\partial y^2} + g\bar{C}\left[P(x, y, t)\right], \tag{8.40}$$

which describes 2D inhomogeneous reaction and diffusion in a dispersive medium. Here $\bar{C}[P(x, y, t)] \equiv \bar{C}(P)$ is a kinetic rate term. We model the reaction kinetics inside the dendrites either by a linear rate term, $\bar{C}(P) = CP$, or by a logistic rate term [145],

$$\bar{C}(P) = CP(x, y, t)[1 - P(x, y, t)]. \tag{8.41}$$

In the latter case, Eq. (8.40) represents a generalization of the FKPP equation [136,137]. Integration with the power law kernel $t^{-\gamma}$ ensures anomalous diffusion in both the dendrite and the spines. Employing the short-hand notation for fractional derivatives, Eq. (8.40) for $P(x, y, t)$ reads,

$$\frac{\partial^\gamma P}{\partial t^\gamma} - \delta(y)I_t^{1-\gamma}\frac{\partial^2 P}{\partial x^2} - g\frac{\partial^2 P}{\partial y^2} = g\bar{C}(P). \tag{8.42}$$

To obtain a first and brief insight into the problem of front propagation, we consider the linear rate term and $\gamma = 1$. In this case, one obtains a "simple" solution [146] for the traveling wave along the x-axis, i.e., along the dendrite. For the marginal probability distribution function, $P_1(x,t) = \int dy P(x,y,t)$, we find

$$P_1(x,t) = \frac{\sqrt{2g^{1/2}}}{\pi \sqrt{t^{1/2}}} \exp\left[\frac{x^2}{2\sqrt{gt}} - Cgt\right]. \tag{8.43}$$

Consequently, the coordinates of the front are $x \sim t^{3/4}$, and the front propagates with the decreasing velocity $v \sim t^{-1/4}$. This solution illustrates the asymptotic failure of the reaction-transport front propagation due to subdiffusion inside spiny dendrites.

8.2.1 *Front propagation in combs: Nonlinear kinetics*

As mentioned above, a mechanism for translocation waves of CaMKII has been suggested in Ref. [131]. Activated CaMKII kinase travels along dendrites with additional translocation inside spines. The process of activation, the conversion of primed CaMKII to its active state, corresponds to an irreversible reaction that, in absence of spines, is described by the FKPP equation [131]. In the framework of the dispersive subdiffusive comb, suggested above (8.42), nonlinear reactions and diffusion in dendrites take place along the x-axis, while subdiffusion in fingers describes the translocation inside the spines. The reaction-transport equation (8.42) now reads

$$\frac{\partial^\gamma P}{\partial t^\gamma} - \delta(y)I_t^{1-\gamma}\left[D\frac{\partial^2 P}{\partial x^2} + \bar{C}(P)\right] = g\frac{\partial^2}{\partial y^2}P. \tag{8.44}$$

Here D describes the diffusivity inside the dendrites, while $\bar{C}(P) = CP(1-P)$ is the nonlinear kinetic rate term. Integrating with respect to y to obtain the kinetic equation for the marginal distribution $P_1(x,t)$, we have

$$\frac{\partial^\gamma P_1}{\partial t^\gamma} - I_t^{1-\gamma}\left[D\frac{\partial^2 f}{\partial x^2} + \bar{C}(f)\right] = 0. \tag{8.45}$$

For the sake of brevity, we have denoted $f = P(x, y = 0, t)$. Consider the fractional comb model (8.44) without reaction,

$$\frac{\partial^\gamma P}{\partial t^\gamma} - \delta(y)I_t^{1-\gamma}D\frac{\partial^2 P}{\partial x^2} = g\frac{\partial^2}{\partial y^2}P, \tag{8.46}$$

and perform the Laplace transformation,

$$s^\gamma \hat{P} - \delta(y)Ds^{\gamma-1}\frac{\partial^2 \hat{P}}{\partial x^2} = g\frac{\partial^2 \hat{P}}{\partial y^2} + s^{\gamma-1}\delta(x)\delta(y), \tag{8.47}$$

with the initial condition $P(t = 0) = \delta(x)\delta(y)$. Looking for the solution in the form

$$\hat{P}(x, y, s) = \exp[-\sqrt{s^\gamma/g}|y|]\hat{f}(x, s), \qquad (8.48)$$

we conclude that $\hat{P}(x, y = 0, s) = \hat{f}(x, s)$, and integration of Eq. (8.48) with respect to y yields

$$\hat{f}(x, s) = \sqrt{\frac{s^\gamma}{4g}}\hat{P}_1(x, s). \qquad (8.49)$$

Substituting (8.49) into (8.45) in Laplace space, we get

$$s\hat{P}_1 - P_1(t = 0) - \frac{Ds^{\gamma/2}}{2\sqrt{g}}\frac{\partial^2 \hat{P}_1}{\partial x^2} - \frac{Cs^{\gamma/2}}{2\sqrt{g}}\hat{P}_1 = -C\mathcal{L}[f^2]. \qquad (8.50)$$

Multiplying this equation by e^{st} and using the identity $e^{st}s^\alpha \hat{f}(s) = \frac{\partial}{\partial t}e^{st}s^{\alpha-1}\hat{f}(s)$, we integrate with the appropriate contour to obtain the inverse Laplace transform. This yields

$$\frac{\partial P_1}{\partial t} - \frac{1}{2\sqrt{g}}\frac{\partial}{\partial t}I_t^{1-\gamma/2}\left[D\frac{\partial^2 P_1}{\partial x^2} + CP_1\right] = -\frac{C}{4g}\left[\frac{\partial}{\partial t}I_t^{1-\gamma/2}P_1\right]^2. \qquad (8.51)$$

The nonlinear term is obtained by the following chain of transformations

$$C[P_0^2] = C\left[\mathcal{L}^{-1}\hat{P}_0\right]^2 = \frac{C}{4g}\left[\mathcal{L}^{-1}s^{\gamma/2}\hat{P}_1\right]^2 = \frac{C}{4g}\left[\frac{\partial}{\partial t}I_t^{1-\gamma/2}P_1\right]^2. \qquad (8.52)$$

8.2.2   *Hyperbolic scaling and the Hamilton–Jacobi technique*

To evaluate the overall velocity of the asymptotic front, we introduce a small parameter, say ε, and carry out a hyperbolic rescaling of time and space [147]: $x \to x/\varepsilon$, $t \to t/\varepsilon$, and $P_1(x, t) \to P_1^\varepsilon(x, t) = P_1(x/\varepsilon, t/\varepsilon)$. We look for the asymptotic solution in the form of the Green approximation,

$$P_1^\varepsilon(x, t) = \exp\left[-\frac{G^\varepsilon(x, t)}{\varepsilon}\right]. \qquad (8.53)$$

The main implication of this procedure is that in the limit $\varepsilon \to 0$, one has $\exp\left[-G^\varepsilon(x, t)/\varepsilon\right] = 0$, except if $G^\varepsilon(x, t) = 0$. This equation determines the position of the propagating reaction front. Moreover, we consider the limit $G(x, t) = \lim_{\varepsilon \to 0} G^\varepsilon(x, t)$ as the principal Hamiltonian function [147], which makes it possible to apply the Hamiltonian approach to the calculation of the front velocity. In this case, the partial derivatives of $G(x, t)$ with respect to time and the coordinate have the physical meaning of the Hamiltonian and momentum:

$$\frac{\partial G(x,t)}{\partial t} = -H \quad \text{and} \quad \frac{\partial G(x,t)}{\partial x} = p. \tag{8.54}$$

Substituting the ansatz (8.53) for the probability distribution function inside the dendrites into Eq. (8.51), we carry out fractional integrations with respect to time. We start with the last term in Eq. (8.51), the kinetic rate term. We rewrite it in the following convenient form,

$$\frac{\varepsilon}{\Gamma(1 - \gamma/2)} \frac{\partial}{\partial t} \int_0^{t/\varepsilon} dt'(t')^{-\gamma/2} \exp[-G^\varepsilon(t - \varepsilon t', x - \varepsilon x')/\varepsilon]. \tag{8.55}$$

Performing the expansion

$$G^\varepsilon(t - \varepsilon t', x - \varepsilon x') \approx G^\varepsilon(x,t) - \varepsilon \frac{\partial G^\varepsilon(x,t)}{\partial t} t' - \varepsilon \frac{\partial G^\varepsilon(x,t)}{\partial x} x', \tag{8.56}$$

and substituting it into Eq. (8.55), we obtain

$$\frac{1}{\Gamma(1 - \gamma/2)} \left[-\frac{\partial G^\varepsilon(x,t)}{\partial t} \right] \exp\left[-\frac{G^\varepsilon(x,t)}{\varepsilon} \right]$$
$$\times \int_0^{t/\varepsilon} (t')^{-\gamma/2} \exp\left[\frac{\partial G^\varepsilon(x,t)}{\partial t} t' + \frac{\partial G^\varepsilon(x,t)}{\partial x} x' \right] dt'. \tag{8.57}$$

Note that we have neglected the differentiation of the upper limit of the integral. This term is of the order of $O(\varepsilon^{1+\gamma/2}) \sim o(\varepsilon)$ and vanishes in the limit $\varepsilon \to 0$. The same procedure of expansion is performed for the diffusion term in Eq. (8.51), which yields

$$\frac{\varepsilon^3}{\Gamma(1 - \gamma/2)} \frac{\partial^3}{\partial t \partial x^2} \exp\left[-\frac{G^\varepsilon(x,t)}{\varepsilon} \right]$$
$$\times \int_0^{t/\varepsilon} (t')^{-\gamma/2} \exp\left[\frac{\partial G^\varepsilon(x,t)}{\partial t} t' + \frac{\partial G^\varepsilon(x,t)}{\partial x} x' \right] dt'. \tag{8.58}$$

Differentiating in the limit $\varepsilon \to 0$ and taking into account that the Hamiltonian H and the momentum p in Eq. (8.54) do not depend explicitly on x and t, which results in the absence of mixed derivatives, we obtain the Laplace transform of the subdiffusive kernel $t^{-\gamma/2}$. The procedures in Eqs. (8.57) and (8.58) turn the kinetic equation (8.51) into a kind of Hamilton–Jacobi equation, which establishes a relation between the Hamiltonian and the momentum,

$$H = \left(\frac{Dp^2 + C}{2\sqrt{g}} \right)^{2/(2-\gamma)}. \tag{8.59}$$

The action is given by $G(x,t) = \int_0^t [p(s)\dot{x}(s) - H(p(s), x(s))] ds$. The rate v at which the front moves is determined by the condition $G(x,t) = 0$. Together with the Hamilton equations, this yields

$$v = \dot{x} = \frac{\partial H}{\partial p}, \quad v = \frac{H}{p}. \tag{8.60}$$

Note that the first equation in (8.60) reflects the dispersion condition, while the second one is a result of the asymptotically free particle dynamics, when the action is $G(x,t) = px - Ht$. Taking into account $x = vt$, we obtain Eq. (8.60). The combination of those two equations can be replaced by

$$v = \min_{H>0} \frac{H}{p(H)} = \min_{p>0} \frac{H(p)}{p}. \tag{8.61}$$

We also have from the front velocity conditions (8.60) that $\partial \ln H / \partial p = 1/p$, which eventually yields from Eq. (8.61),

$$v = \left[\left(\frac{4}{g} \right)^{2/(2-\gamma)} \frac{D}{2-\gamma} \left(\frac{C}{2+\gamma} \right)^{(2+\gamma)/(2-\gamma)} \right]^{1/2}. \tag{8.62}$$

To proceed, we note first that the limiting case of this result with $\gamma = 0$ corresponds to CaMKII propagation along the dendrite only, i.e., no fingers/teeth are present. Equation (8.62), after rescaling $D/\sqrt{g} \to D$ and $C/\sqrt{g} \to C$, recovers the FKPP scheme for $\gamma = 0$, which yields $v = \sqrt{DC}$.

Note also that no propagation failure of the activation front occurs. This result is due to the absence of a kinetic "sink" term, $-k_t P$, in Eq. (8.44), i.e., due to neglecting the possibility that the spines accumulate a large amount of Ca^{2+} [133, 134], where k_t is a translocation/accumulation rate [131]. Including this term in Eq. (8.44), we anticipate that the hyperbolic scaling for this new equation will yield a solution similar to Eq. (8.61) with $H = 0$, which corresponds to propagation failure.

Adding the sink term in Eq. (8.44), we obtain

$$\frac{\partial^\gamma P}{\partial t^\gamma} - \delta(y) I_t^{1-\gamma} \left[D \frac{\partial^2 P}{\partial x^2} + CP(1-P) \right] - g \frac{\partial^2 P}{\partial y^2} - gk_t P = 0. \tag{8.63}$$

Repeating the same procedures of the Laplace transformation and integration with respect to y, with $\hat{P}_1 = \int_{-\infty}^{\infty} \hat{P}(x,y,s) dy$, and using the substitution

$$P_1(x,s) = 2\sqrt{g/s^\gamma} P(x, y=0, s), \tag{8.64}$$

we find

$$s\hat{P}_1 - \delta(x) = \frac{D\sqrt{s^\gamma}}{2\sqrt{g}} \frac{\partial \hat{P}_1}{\partial x^2} + \frac{C\sqrt{s^\gamma}}{2\sqrt{g}} \hat{P}_1 - k_t g s^{1-\gamma} \hat{P}_1. \tag{8.65}$$

Here we neglect the nonlinear term, which does not contribute to the Hamilton–Jacobi equation in the hyperbolic scaling approximation. Again multiplying this equation by e^{st} and using the same identity $e^{st}s^{\alpha}f(s) = \frac{\partial}{\partial t}e^{st}s^{\alpha-1}f(s)$, as above, we obtain the inverse Laplace transform:

$$\frac{\partial P_1}{\partial t} = \frac{D}{2\sqrt{g}}\frac{\partial}{\partial t}I_t^{1-\gamma/2}\left[\frac{\partial P_1}{\partial x^2} + \frac{C}{D}P_1\right] - k_t g\frac{\partial}{\partial t}I_t^{\gamma}P_1. \tag{8.66}$$

Application of the hyperbolic scaling with the asymptotic solution (8.53) yields

$$2\sqrt{g}H = Dp^2 H^{\gamma/2} + CH^{\gamma/2} - 2k_t g^{3/2}H^{1-\gamma}. \tag{8.67}$$

Let us consider a specific case, namely $\gamma = 2/3$, which yields

$$H = \left(\frac{Dp^2 + C - 2k_t g^{3/2}}{2\sqrt{g}}\right)^{3/2}. \tag{8.68}$$

For $C > 2k_t g^{3/2}$, no propagation failure occurs and the front moves asymptotically with a constant velocity. For $C \leq 2k_t g^{3/2}$, the only solution is $H = 0$, which yields $v = 0$. In other words, $2k_t g^{3/2}$ is the minimum reaction rate necessary to sustain propagation along the spiny dendrite due to the presence of translocation into spines. On the other hand, $(C/2k_t)^{2/3}$ can be viewed as the minimum value for the density of spines necessary for propagation failure to occur. Both results are in agreement with the results obtained from very different models based on the cable model [135].

In the general case, we compare the competition between the activation term $CH^{\gamma/2}$ and the translocation term $-2k_t g^{3/2}H^{1-\gamma}$ in Eq. (8.67) in the limit $H \to 0$. For $\gamma \in [2/3, 1)$, the translocation term is dominant and leads to the solution with $H = 0$ and propagation failure. For $0 < \gamma < 2/3$, the activation in dendrites can be dominant. This situation is more complicated, and the activation-translocation front can propagate with an asymptotically finite velocity.

8.2.3 *Front propagation in combs: Linear kinetics*

Finally, we consider the linear counterpart of Eq. (8.44) with the kinetic term $\bar{C}(P) = CP$. This analysis allows us to elucidate the behavior of the tail of the marginal distribution $P_1(x,t)$ and to determine if the front accelerates or travels with constant velocity. As follows from the fractional differentiation of Eq. (8.51), the equation for $P_1(x,t)$ reads, see also [16],

$$\frac{\partial^{1-\gamma/2}P_1}{\partial t^{1-\gamma/2}} = \frac{D}{2\sqrt{g}}\left[\frac{\partial^2}{\partial x^2} + \frac{C}{D}\right]P_1, \tag{8.69}$$

with the initial condition $P_1(x, t = 0) = \delta(x)$. After Fourier transformation, $\mathcal{F}[P_1(x,t)] = \tilde{P}_1(k,t)$, we obtain the solution in the form of the Mittag–Leffler function,

$$\tilde{P}_1(k,t) = E_{1-\gamma/2}\left[A(k)t^{1-\gamma/2}\right], \tag{8.70}$$

where $A(k) = (C - Dk^2)/2\sqrt{g}$. In the asymptotic regime, when $x, t \gg 1$, we have $C \gg Dk^2$. Using the asymptotic behavior of the Mittag–Leffler function for large positive arguments [43], we find

$$\tilde{P}_1(k,t) \approx \exp\left[\left(\frac{C}{2\sqrt{g}} - \frac{D}{2\sqrt{g}}k^2\right)^{2/(2-\gamma)} t\right]$$

$$\approx \exp\left[\left(\frac{C}{2\sqrt{g}}\right)^{2/(2-\gamma)}\left(1 - \frac{2}{2-\gamma}\frac{Dk^2}{C}\right)t\right]. \tag{8.71}$$

Applying the inverse Fourier transformation, we obtain

$$P_1(x,t) = \exp\left[\left(\frac{C}{2\sqrt{g}}\right)^{2/(2-\gamma)} t - \frac{(2-\gamma)x^2(2\sqrt{g})^{2(2-\gamma)}}{8DC^{\gamma/(2-\gamma)}t}\right], \tag{8.72}$$

which implies that the reaction front propagates with a nonzero and constant overall velocity. Note that for normal diffusion, $\gamma = 0$, one obtains the Fisher velocity $v = \sqrt{DC/g} \to \sqrt{DC}$, see the limiting case $\gamma = 0$ in Eq. (8.62).

8.2.4   *Conclusion*

We have considered a comb as a convenient model for the analytical exploration of anomalous transport and front propagation phenomena along spiny dendrites. The physical arguments provided above explain why anomalous transport, namely subdiffusion, of either CaMKII or neutral particles is possible and support the use of the comb model. These arguments are based on the geometry of dendritic spines, which determines the expression for the transport exponent in Eq. (8.5). This situation becomes more subtle in the case of the nonlinear FKPP reaction. As shown above, the power law kernel of the transition probability based on geometry arguments is insensitive to the nonlinear kinetic term. This result differs completely from a mesoscopic non-Markovian approach, developed in [148,83], where the spines-dendrite interaction and an extension including reactions in spines were described in the framework of a variable residence time. In such a model, the nonlinear kinetic term leads to complications in the transition probability [149,83].

In the next chapter we present some additional arguments to justify the reaction-transport theory discussed in the chapter.

Chapter 9

Front propagation: Actin polymerization in a comb micrograph

9.1 Introduction

Microfluidics has become an indispensable tool of modern biophysical research. It provide a means to perform complex single-cell experiments with a huge throughput and a high level of control. Custom geometries and control of flows and chemical reactions can be realized due the flexible design of microfluidic experiments. For example, a microfluidic device with a main supply channel and numerous identical side channels or chambers of different shapes, as shown in Fig. 9.1, has recently been employed to probe the dynamics of actin polymerization [150, 151]. The main channel serves to deliver reagents to the side chambers, where the corresponding reactions can be observed. The flow in the main channel and diffusion in the side-channels are the dominant means of transport in such devices. This experimental device for the study of actin polymerization [150, 152] is a direct application of the comb model, and the effects of complex diffusion should play a substantial role in the observed phenomena.

The goal of this chapter is to combine the tools of anomalous transport and reaction dynamics, developed in the previous chapters, to provide a theoretical foundation for the corresponding experimental efforts [128]. Analytical results on reaction propagation can guide the design of microfluidic devices and also lead to experimental tests of anomalous diffusion and reaction dynamics. For the polymerization reaction we observe, depending on the reactant concentrations, either propagation failure of the reaction front [153, 154], or a front with finite speed, which eventually leads to the FKPP velocity in the long-time asymptotic regime [136, 137]. These two processes correspond to different time scales in the finite comb geometry of the experimental device. While the FKPP velocity occurs in the long-time

141

asymptotic regime, the subdiffusive reaction front propagation corresponds to a transient process and takes place at an intermediate time scale. A rigorous derivation of the governing equations allowes us to calculate explicitly the characteristic time scale separating these two regimes. This time scale is determined by the geometry of the microfluidic device and can be used to tune the regimes of diffusion and reactions in the experiments.

9.1.1 *Experimental setup*

A multi-height microfluidic device has been used to study the dynamics of actin polymerization in diffusion-controlled comb-like structures [150]. The microfluidic system consists of a main channel with a width $h_1 = 5\mu$m, connected to comb-like structures that are smaller in width, $h_2 = 0.5\mu$m. Since the height of the main channel is ten-fold larger than that of the comb-like structures, a diffusive interface between the main advective channel and the comb-like structure is generated due to the large hydraulic resistance of the connecting structures. This makes it possible to add a solution of polymerization-inducing potassium chloride (KCl) to a solution of monomeric fluorescently labeled actin, including the ATP necessary for in vitro polymerization, via the main channel. The reagent KCl diffuses into the comb-like structure and induces the polymerization of actin monomers into filaments. Similarly, magnesium ion (Mg^{2+}) can be used to induce the assembly of actin filaments into fibers. A general reaction scheme of the actin polymerization with magnesium ion or KCl as an inducer has been suggested, and was studied, in Ref. [128]. In experiments, the design of the side chambers can be varied; they can, for example, have circular or rectangular shapes. These shapes can also be incorporated into the analytical approach, considered in the next section.

9.2 Mapping the Laplace operator onto a comb equation

Mapping the Laplace operator acting in a three dimensional continuous-discrete geometry, as in Fig. 9.1, onto a continuous two dimensional comb model geometry is related to averaging in x-y-z-space over some characteristic volume. Our goal is to describe anomalous diffusion of the inducer on the comb by the two dimensional pdf $P(x, y, t)$. In this case, the process of mapping of the Laplace operator $\Delta P(x, y, z)$ onto the comb model corresponds to establishing relations between the parameters of the geometrical

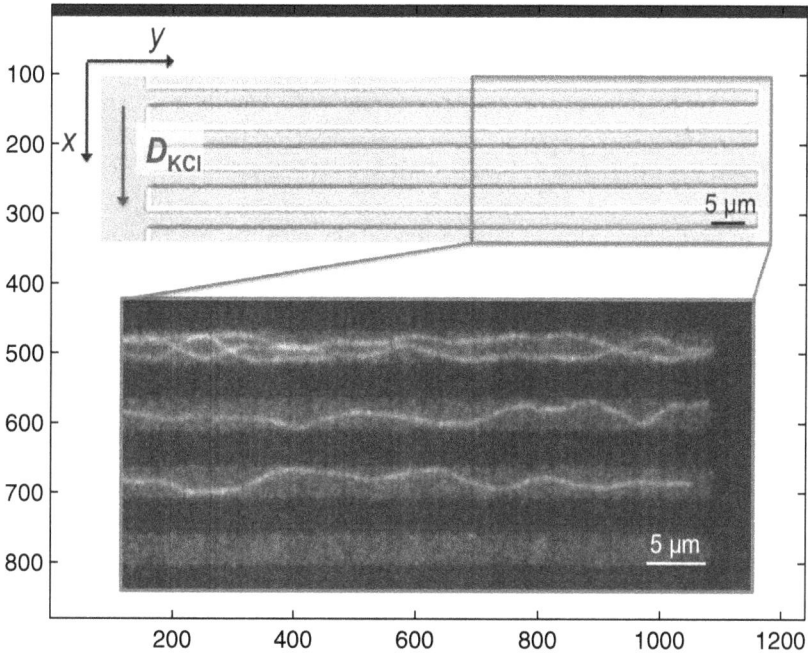

Fig. 9.1   Top: Optical micrograph of a segment of the microfluidics comb-like structures. Bottom: Microfluidic micrographs of fluorescently-labeled, polymerized actin filaments in a comb-like structure. Reprinted from Chaos, Solitons & Fractals, 92, A. Iomin, V. Zaburdaev, and T. Pfohl, Reaction front propagation of actin polymerization in a comb-reaction system, 115–122 (2016) with permission from Elsevier.

constraints on the Laplace operator and the transport constants D_x and D_y.

In the description of the microfluidic device, the Laplace operator acts on the distribution functions in the bulk of the main channel, $P(x,y,z) = P_b(x,y,z)$, and in the fingers (the side channels, where reactions take place), $P(x,y,z) = P_f(x,y,z)$. This suggests the following mapping procedure. In the bulk, with infinite extension along the x-coordinate and limited by a yz-surface with a cross-section $a \times a$ along the perpendicular coordinates, the Laplace operator is given by

$$D\Delta P_b(x,y,z) = D(\partial_x^2 + \partial_y^2 + \partial_z^2)P(x,y,z). \qquad (9.1)$$

The diffusivity of the inducer is denoted by D, and the boundary conditions for $P(x, y, z) = P_b(x, y, z)$ are:

$$\partial_z P|_{z=-a/2} = \partial_z P|_{z=a/2} = \partial_y P|_{y=-a/2} = \partial_y P|_{y=a/2} = 0. \qquad (9.2)$$

Integration with respect to z in Eq. (9.1) removes the z-component of the Laplace operator due to the boundary condition. For the pdf we obtain

$$\int_{-a/2}^{a/2} P_b(x, y, z)dz \approx aP_b(x, y, 0), \qquad (9.3)$$

where we used the midpoint rule. Integration with respect to y in the bulk yields zero, except for those y-regions where the bulk is connected to the fingers. To obtain the dynamics along the backbone, which corresponds to $y = 0$, we "plunge" the finders into the bulk. From a mathematical point of view, this means applying again the midpoint rule. We have for the bulk diffusion at arbitrary x,

$$\frac{1}{a^3} \int_{-a/2}^{a/2} dxdydz\Delta P_b(x, y, z) \approx D\partial_x^2 P(x, y = 0, z = 0) = D\delta(y/a)\partial_x^2 P(x, y). \qquad (9.4)$$

We have omitted the z-coordinate in the distribution function, $P(x, y, z = 0) \equiv P(x, y)$.

Next we consider the fingers, which have length h and their x-z-cross-section is $b \times b$. To work with a symmetrical pdf, we map the Laplace operator onto a two-sided symmetrical comb model, reflected in the choice of the symmetric boundary conditions at $y = \pm h$. The Laplace operator with diffusivity d inside the fingers reads

$$d\Delta P_f(x, y, z) = d(\partial_x^2 + \partial_y^2 + \partial_z^2)P(x, y, z). \qquad (9.5)$$

Taking into account the boundary conditions, integration/averaging with respect to x and z results in zero, except for the regions of the fingers, which are periodic in x, where we obtain ∂_y^2. We have for a single finger

$$\frac{1}{b^3} \int_{-b/2}^{b/2} dxdydz\partial_y^2 P(x, y, z) \approx \partial_y^2 P(x, y, z = 0). \qquad (9.6)$$

If the fingers have a density ρ, then the average Laplace operator for the fingers is given by

$$\frac{d}{b^3} \int_{-b/2}^{b/2} dxdydz \, d\Delta P_f(x, y, z) \approx \rho d\partial_y^2 P(x, y, z = 0). \qquad (9.7)$$

The finger density ρ is the number of fingers in an interval of length a along the x-direction. Without restriction of generality we set $\rho \sim a/b$. Since the

z-component disappears from the averaged Laplace operator, we omit z again in the distribution function, $P(x, y, z = 0) \equiv P(x, y)$.

After combining the equations (9.4) and (9.7), we obtain an effective equation for the inducer transport along the comb structure with parameters related to the experimental geometry, a and b, and diffusion constants D and d,

$$\partial_t P = aD\delta(y)\partial_x^2 P + d_\rho \partial_y^2 P. \tag{9.8}$$

The rescaled diffusion coefficient $d_\rho = d\rho = D_y$ and $aD = D_x$ establish a relation with the diffusion tensor of the comb model [15]. Note that the diffusion coefficient in the x-direction is $D(y) = aD\delta(y) = D\delta(y/a)$. The scaling of the δ function by a is convenient and valid in the limit $h \to \infty$.

We conclude this section with an important comment. The singularity of the diffusion coefficient results from the mapping of the Laplace operator onto the comb equation (9.8); it is the defining feature of the comb model. This singularity of the diffusion coefficient is not trivial and relates to a non-zero flux along the x-coordinate. Integrating Eq. (9.8) with respect to y from $-\epsilon/2$ to $\epsilon/2$, $\int_{-\epsilon/2}^{\epsilon/2} dy \cdots$, we obtain for the lhs of the equation, by applying the midpoint rule, $\epsilon \partial_t P(x, y = 0, t)$, which is exact in the limit $\epsilon \to 0$. This term can be neglected in the limit $\epsilon \to 0$. Considering the rhs of the equation, we find that the term responsible for the transport in the y-direction reads

$$d_\rho \partial_y \left[P(x, y, t)\big|_{y=\epsilon/2} - P(x, y, t)\big|_{y=-\epsilon/2} \right]. \tag{9.9}$$

This corresponds to the two outgoing fluxes from the backbone in the $\pm y$ directions, $F_{+y} + F_{-y}$. The transport in the x-direction, after integration, is

$$\epsilon D(y \to 0)\partial_x^2 P(x, y = 0, t) = F_{x;y=0} \equiv F_x. \tag{9.10}$$

Here we consider a general diffusivity function in the x-direction, $D(y)$, instead of the specific form $aD\delta(y)$ in Eq. (9.8). The second derivative with respect to x can be written in the form $\epsilon \partial_x^2 P = [\partial_x P(x+\epsilon/2) - \partial_x P(x-\epsilon/2)]$, which shows explicitly that it accounts for both incoming and outgoing contributions to the flux F_x along the x-direction at point x. Kirchhoff's law implies that $F_x + F_{+y} + F_{-y} = 0$ for every point x at $y = 0$. As discussed above, the function F_x contains both incoming and outgoing fluxes of the probability, whereas F_{+y} and F_{-y} are both outgoing probability fluxes. If the latter outgoing fluxes are not zero, then the flux F_x has to be nonzero as well, $F_x \neq 0$; it must contain an incoming flux. Therefore, $\epsilon D(y \to 0) \neq 0$.

Taking $D(y) = \epsilon a D[\pi(y^2 + \epsilon^2)]$, we obtain in the limit $\epsilon \to 0$ a nonzero flux F_x with $D(y) = Da\delta(y)$, which is the diffusion coefficient in the x-direction in Eq. (9.8). It is worth noting that this transport property of the comb model makes it possible to map the general Laplace operator of the experimental setup onto the comb model by ensuring an "isomorphism" of the geometry or topology of the probability/particle fluxes.

The experimental setup imposes the following boundary conditions on Eq. (9.8), $P(x = \pm\infty, y, t) = 0$, $\partial_x P(x, y, t)|_{x=\pm\infty} = 0$, and $\partial_y P(x, y, t)|_{y=\pm h} = 0$. We impose the initial condition $P(x, y, t = 0) = P_0(x)\delta(y)$. Such a localized initial condition and the absence of the convective flows can be achieved experimentally.

9.3 Reaction-transport in a comb

We augment the comb equation (9.8) with a reaction term $\rho\bar{C}(P)$, which also accounts for the density of fingers,

$$\frac{\partial P}{\partial t} = D\delta(y/a)\frac{\partial^2 P}{\partial x^2} + d_\rho\frac{\partial^2 P}{\partial y^2} - \rho\bar{C}(P). \tag{9.11}$$

Equation (9.11) describes anomalous diffusion of an inducer in a comb structure together with the actin polymerization reaction. In the simplest case, the reaction term $C(P)$ is given by the second order kinetic equation:

$$\frac{dP}{dt} = -\kappa(N - P_0 + P)P, \tag{9.12}$$

where N is the initial concentration of actin, while κ is the rate constant of the reaction.

9.3.1 *Auxiliary: The reaction term*

The reaction term $\bar{C}(P)$ can be found as follows, see for example Ref. [155]. We describe the polymerization reaction by the chemical equation,

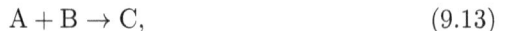

$$A + B \to C, \tag{9.13}$$

where [A] is the concentration of actin, [B] the concentration of the inducer, and [C] the concentration of the polymer. The inducer is magnesium ion or KCl. We assume that the polymerization reaction (9.13) obeys the Law of Mass Action,

$$\frac{dP}{dt} = -\kappa P \cdot [A], \tag{9.14}$$

where $P(x, y, t) = [B]$, κ is the rate constant, and the initial conditions are $[B(t = 0)] \equiv P(t = 0) = P_0$, $[A(t = 0)] = [A]_0 = N$, and $[C]_0 = 0$.

Taking into account the stoichiometry of (9.13) and the conservation of the total amount of actin, we can express $[A]$ in terms P:

$$P \equiv [B] = [B]_0 - [C] = P_0 - [C], \tag{9.15}$$
$$[A] = [A]_0 - [C] = [A]_0 - P_0 + P. \tag{9.16}$$

Substituting (9.16) into (9.14) and recalling that $[A]_0 = N$, we find

$$\frac{dP}{dt} = -\kappa(N - P_0 + P)P. \tag{9.17}$$

9.3.2 Solution of the linear reaction-transport comb equation: Extinction dynamics

If the reaction occurs under conditions such that $N - P_0 \gg P$ and $N \gg P_0$, i.e., there is much more of actin than inducer, then the second order kinetic equation (9.12) can be simplified to a first order equation,

$$\frac{dP}{dt} = -\kappa N P. \tag{9.18}$$

Substitution of Eq. (9.18) into Eq. (9.11) yields the reaction-transport equation

$$\frac{\partial P}{\partial t} = Da\delta(y)\frac{\partial^2 P}{\partial x^2} + d_\rho\frac{\partial^2 P}{\partial y^2} - CP, \tag{9.19}$$

where $C = \rho\kappa N$ is the constant rate coefficient.

We remove the reaction term from Eq. (9.19) via the transformation

$$P = e^{-Ct}\bar{P}. \tag{9.20}$$

The next step is a Laplace transformation, which turns Eq. (9.19) into

$$s\hat{P} = Da\delta(y)\frac{\partial^2 \hat{P}}{\partial x^2} + d_\rho\frac{\partial^2 \hat{P}}{\partial y^2} + P_0\delta(y). \tag{9.21}$$

We assume that the solution can be written as a product $\hat{P}(x, y, s) = \hat{n}(y, s)f(x, s)$. The solution inside the fingers is found from the equation in Laplace space,

$$s\hat{n}(y) = d_\rho\frac{\partial^2 \hat{n}(y)}{\partial y^2}, \tag{9.22}$$

with the boundary condition $\partial_y\hat{n}(y)|_{y=\pm h} = 0$, which yields

$$\tilde{n}(y, s) = \frac{\cosh\left[(h - |y|)\sqrt{s/d_\rho}\right]}{\cosh\left[h\sqrt{s/d_\rho}\right]}, \tag{9.23}$$

and $n(y = 0) = 1$.

Since we seek the solution in the form

$$\hat{P}(x, y, s) = \hat{n}(y, s) f(x, s), \tag{9.24}$$

we can write its second derivative with respect to y as

$$\frac{\partial^2 \hat{P}}{\partial y^2} = \left\{ -2\delta(y)\sqrt{s/d_\rho} \frac{-\sinh\left[(h - |y|)\sqrt{s/d_\rho}\right]}{\cosh\left[h\sqrt{s/d_\rho}\right]} + [s/d_\rho]\hat{n}(y) \right\} f(x, s). \tag{9.25}$$

Therefore the diffusion of the inducer in the bulk-backbone, determined by $f(x, s)$, is described by the following equation,

$$Da\partial_x^2 f - 2\sqrt{sd_\rho}\tanh\left[h\sqrt{s/d_\rho}\right]f + P_0 = 0. \tag{9.26}$$

The presence of the hyperbolic tangent function implies the existence of two time scales and their corresponding dynamic regimes. On a short time scale, when $h\sqrt{s/d_\rho} \gg 1$, we have

$$\tanh[h\sqrt{s/d_\rho}] \approx 1, \tag{9.27}$$

and

$$D_{\frac{1}{2}}\partial_x^2 f - \sqrt{s}f + f_0 = 0. \tag{9.28}$$

Here

$$D_{\frac{1}{2}} = \frac{Da}{2\sqrt{d_\rho}} = \frac{D\sqrt{ab}}{2\sqrt{d}} \tag{9.29}$$

is a generalized diffusion coefficient,[1] and $f_0(x) = P_0(x) = P_0\delta(x)$, for simplicity. The generalized diffusion coefficient in the x-direction is independent of the finger length h.

By performing the Fourier transformation $\tilde{f}(k) = \mathcal{F}[f(x)]$, we obtain

$$\tilde{f}(k, s) = \frac{P_0}{\sqrt{s} + D_{\frac{1}{2}}k^2}. \tag{9.30}$$

Carrying out the inverse Laplace transformation of the solution (9.30) and taking into account the definition of the Mittag–Leffler function, we arrive at the solution

$$\bar{f}(k, t) = \frac{P_0}{2\pi i} \int_{\sigma - i\infty}^{\sigma + i\infty} \frac{e^{st}ds}{\sqrt{s} + D_{\frac{1}{2}}k^2} = \frac{P_0}{\sqrt{t}} E_{\frac{1}{2}, \frac{1}{2}}\left(-D_{\frac{1}{2}}k^2 t^{\frac{1}{2}}\right). \tag{9.31}$$

---

[1] Note that $\rho = a/b$.

The inverse Fourier transform of Eq. (9.31) yields a solution in terms of the Fox function. This solution is valid for the "short" time-scale $t \ll h^2/d_\rho = h^2 b/da$. It is instructive to obtain an approximation of the solution in terms of analytical functions. Using the property $E_{a,b}(z) = zE_{a,a+b}(z) + 1/\Gamma(b)$ for the two parameter Mittag–Leffler functions [156], and taking into account that the Mittag–Leffler function $E_{a,1}(z)$ for small arguments, $|z| \ll 1$ and $z < 0$, can be approximated by exponentials [43], we obtain

$$E_{\frac{1}{2},\frac{1}{2}}\left(-D_{\frac{1}{2}}k^2t^{\frac{1}{2}}\right) \approx -D_{\frac{1}{2}}k^2t^{\frac{1}{2}}\exp\left[-\frac{D_{\frac{1}{2}}k^2t^{\frac{1}{2}}}{\Gamma(3/2)}\right] + \frac{1}{\Gamma(1/2)}, \qquad (9.32)$$

where $\Gamma(3/2) = (1/2)\Gamma(1/2) = \sqrt{\pi}/2$, see Eq. (2.2) and Eq. (2.4). Now the inverse Fourier transformation can be easily performed and yields for the approximation

$$\left[\frac{\Gamma(3/2)}{4\pi D_{\frac{1}{2}}t^{\frac{1}{2}}}\right]^{\frac{1}{2}}\frac{d^2}{dx^2}\exp\left[-\frac{x^2\Gamma(3/2)}{4D_{\frac{1}{2}}\sqrt{t}}\right]. \qquad (9.33)$$

Taking into account the reaction term, we obtain the short-time scale solution for diffusion in the bulk,

$$P(x,y=0,t \ll t_0) = \frac{P_0 e^{-Ct}\delta(x)}{\sqrt{\pi t}} + \frac{P_0 e^{-Ct}D_{\frac{1}{2}}^{\frac{1}{2}}}{2(4\pi t)^{\frac{1}{4}}}\frac{d^2}{dx^2}\exp\left(-\frac{\sqrt{\pi}x^2}{8D_{\frac{1}{2}}\sqrt{t}}\right). \qquad (9.34)$$

This result is valid for the time scale $t \ll t_0$, where the time parameter

$$t_0 = \frac{h^2}{d_\rho} = \frac{h^2 b}{ad}, \qquad (9.35)$$

is determined by the geometry of the experimental device.[2] Since Eq. (9.34) is independent of h, it remains valid in the limit $h \to \infty$.

On a long time-scale, the hyperbolic tangent in Eq. (9.26) has a small argument,

$$\tanh[h\sqrt{s/d_\rho}] \approx h\sqrt{s/d_\rho}. \qquad (9.36)$$

This corresponds to an equation for normal diffusion,

$$D(a/2h)\partial_x^2 f - sf + P_0/h = 0, \qquad (9.37)$$

with a well-known solution. The long-time scale solution in the bulk reads

$$P(x,y=0,t \gg t_0) = e^{-Ct}\int dx' P_0(x')\frac{1}{\sqrt{\pi h^2 \bar{D}t}}\exp\left(-\frac{(x-x')^2}{4\bar{D}t}\right), \qquad (9.38)$$

---

[2]Note that if $\rho = 1$, then $t_0 = h^2/d$, which coincides with the transient time of the "classical" comb model [119].

where $\bar{D} \equiv D(a/2h)$. When $P_0(x) = \delta(x)$, the Green function coincides with the distribution.

In Eqs. (9.34) and (9.38), we have seen that there exists a distinct time scale separating the transport regimes, namely t_0. Note that depending on the parameter values, the initial, subdiffusive dynamics may be the only experimentally detectable regime due to the exponential pre-factor e^{-Ct}.

9.4 Reaction front propagation in a case of a high inducer concentration

When the concentration of the inducer is sufficiently high, the approximate Eq. (9.18) is not valid, and we have to work with the second order kinetics, i.e., Eq. (9.12). We write the reaction term as

$$C(P) = CP + C_1 P^2, \quad C = \rho \kappa (N - P_0), \quad C_1 = \rho \kappa. \tag{9.39}$$

In this nonlinear case, the exact analytical treatment of Eq. (9.11) is impossible, and we apply the analytical approximation of Sec. 8.2.2 to find the overall velocity of the polymerization reaction front without determining the exact shape of the front, namely without knowledge of the exact distribution function $P(x, y, t)$.

Since we are interested in the front propagation in the x-direction, the exact shape in the y direction is not important. We consider the distribution as a function of the x-coordinate only, $P_1(x, t)$. In other words, the details of the dynamics inside the fingers is not important, and we take into account only its overall contribution to diffusion along the x-axis. To this end, we integrate with respect to the y-coordinate to obtain the marginal pdf,

$$P_1(x, t) = \int_{-h}^{h} P(x, y, t) dy. \tag{9.40}$$

Such an integration of Eq. (9.11), with the reaction term given by Eq. (9.39), yields an equation that is not closed. Taking into account that

$$\int_{-h}^{h} \partial_y^2 P(x, y, t) dy = \partial_y P(x, y, t) \Big|_{y=-h}^{y=h} = 0, \tag{9.41}$$

we obtain

$$\frac{\partial P_1(x, t)}{\partial t} = aD \frac{\partial^2 P(x, y = 0, t)}{\partial x^2} - CP_1 - C_1 \int_{-h}^{h} P^2(x, y, t) dy. \tag{9.42}$$

We encounter two problematic terms. The first one is $P(x, y = 0, t)$, which will be expressed via $P_1(x, t)$ as done in previous chapters. However, here we need to take into account the finite boundaries. To this end, we use the relation (9.24), noting that $\hat{P}(x, y = 0, s) = f(x, s)$. Integration of Eq. (9.24) with respect to y yields

$$\hat{P}_1(x, s) = \hat{P}(x, y = 0, s) \int_{-h}^{h} \hat{n}(y, s)dy = 2\hat{P}(x, y = 0, s) \int_{0}^{h} \hat{n}(y, s)dy.$$
$$(9.43)$$

Therefore we can write

$$\hat{P}(x, y = 0, s) = \sqrt{s/4d_\rho} \tanh^{-1}\left[h\sqrt{s/d_\rho}\right] \cdot \hat{P}_1.$$
$$(9.44)$$

Performing the inverse Laplace transformation, we obtain from Eq. (9.44),

$$\begin{aligned} P(x, y = 0, t) &= \frac{1}{2\pi i} \int_{\sigma-i\infty}^{\sigma+i\infty} ds \int_{0}^{\infty} dt_1 dt_2 \, P_1(x, t_1) \mathcal{R}_1(t_2) e^{-st_1} e^{-st_2} e^{st} \\ &= \int_{0}^{\infty} dt_1 \int_{0}^{\infty} dt_2 P_1(t_1) \mathcal{R}_1(t_2) \delta(t - t_1 - t_2) \\ &= \int_{0}^{t} dt' P_1(t - t') \mathcal{R}_1(t'). \end{aligned}$$
$$(9.45)$$

Here we take into account the causality principle, $P(t - t_1) = 0$ for $t_1 > t$, and define the kernel $\mathcal{R}_1(t)$ via the inverse Laplace transformation,

$$\mathcal{R}_1(t) = \mathcal{L}^{-1}\left[\sqrt{s/4d_\rho} \tanh^{-1}\left(h\sqrt{s/d_\rho}\right)\right].$$
$$(9.46)$$

The second problematic term is the integration of the nonlinear reaction term, which we call NLRT. We present it in a form convenient for further analytical treatment.

9.4.1 *Chain of transformation for NLRT*

We have the following chain of transformations for the nonlinear reaction term

$$\text{NLRT} \equiv \int_{-h}^{h} P^2(x, y, t)dy.$$
$$(9.47)$$

Defining the pdf $P(x, y, t)$ by the inverse Laplace transformation in time we have

$$(2\pi i)^2 \cdot (\text{NLRT})$$
$$= \int_{\sigma-i\infty}^{\sigma+i\infty} e^{s_1 t} ds_1 e^{s_2 t} ds_2 \int_{-h}^{h} \hat{P}(x, y, s_1) \hat{P}(x, y, s_2) dy$$

$$= \int_{\sigma-i\infty}^{\sigma+i\infty} e^{s_1 t} ds_1 e^{s_2 t} ds_2 \hat{P}(x, y = 0, s_1) \hat{P}(x, y = 0, s_2)$$

$$\times \int_{-h}^{h} \hat{n}(y, s_1) \hat{n}(s_2, y) dy. \tag{9.48}$$

Taking into account the solution in Eq. (9.44), we have

$$(2\pi i)^2 \cdot (\text{NLRT}) = \int_{\sigma-i\infty}^{\sigma+i\infty} ds_1 \, ds_2 \hat{\mathcal{R}}(s_1, s_2) \hat{P}_1(x, s_1) \hat{P}_1(x, s_2) \frac{\sqrt{s_1 s_2}}{4d_\rho}$$

$$\times \tanh^{-1}\left[h\sqrt{s_1/d_\rho}\right] \cdot \tanh^{-1}\left[h\sqrt{s_2/d_\rho}\right] e^{s_1 t} e^{s_2 t}$$

$$= \int_0^\infty d\tau_1 d\tau_2 P_1(x, \tau_1) P_1(x, \tau_2) \int_{\sigma-i\infty}^{\sigma+i\infty} ds_1 \, ds_2$$

$$\times \frac{\sqrt{s_1 s_2}}{4d_\rho} \cdot \tanh^{-1}\left[h\sqrt{s_1/d_\rho}\right] \cdot \tanh^{-1}\left[h\sqrt{s_2/d_\rho}\right]$$

$$\times e^{s_1(t-\tau_1)} e^{s_2(t-\tau_2)} \hat{\mathcal{R}}(s_1, s_2), \tag{9.49}$$

where

$$\hat{\mathcal{R}}(s_1, s_2) = \int_{-h}^{h} \hat{n}(y, s_1) \hat{n}(s_2, y) dy. \tag{9.50}$$

Introducing the new variables $t - \tau_1 = t_1$ and $t - \tau_2 = t_2$, we obtain for the time integration

$$\int_0^\infty d\tau_1 = -\int_t^{-\infty} dt_1 = \int_{-\infty}^t dt_1, \tag{9.51}$$

and the same procedure is performed for τ_2.

Performing this chain of transformations, one eventually arrives at the following equation,

$$\text{NLRT} = \int_0^\infty dt_1 dt_2 P_1(x, t - t_1) P_1(x, t - t_2)$$

$$\times \frac{1}{(2\pi i)^2} \int_{\sigma-i\infty}^{\sigma+i\infty} \hat{\mathcal{R}}(s_1, s_2) e^{s_1 t_1} e^{s_2 t_2} ds_1 \, ds_2. \tag{9.52}$$

Finally, the reaction-transport equation for $P_1(x, t)$ reads

$$\frac{\partial P_1}{\partial t} = aD \frac{\partial^2}{\partial x^2} \int_0^t \mathcal{R}_1(t') P_1(t - t') dt' - CP_1$$

$$- C_1 \int_{-\infty}^t \mathcal{R}(t', t'') P_1(t - t') P_1(t - t'') dt' dt'', \tag{9.53}$$

where the kernel $\mathcal{R}(t', t'')$ is defined in Laplace space, Eq. (9.50).

9.5 Hyperbolic scaling for the overall velocity of the propagating reaction front

To evaluate the overall velocity of the asymptotic front, we follow the hyperbolic scaling approach discussed in Sec. 8.2.2. Following that procedure, we introduce a small parameter, ε, and we rescale $x \to x/\varepsilon$ and $t \to t/\varepsilon$, while for the pdf we have

$$P_1(x,t) \to P_1^\varepsilon(x,t) = P_1\left(\frac{x}{\varepsilon}, \frac{t}{\varepsilon}\right). \tag{9.54}$$

We look for the asymptotic solution in the form of the Green's approximation,

$$P_1^\varepsilon(x,t) = \exp\left[-\frac{G^\varepsilon(x,t)}{\varepsilon}\right]. \tag{9.55}$$

As discussed in Chapter 8, the main point of this approach is that the limit $\varepsilon \to 0$ yields the asymptotic behavior for finite x and t, where

$$\exp\left[-\frac{G^\varepsilon(x,t)}{\varepsilon}\right] = 0, \tag{9.56}$$

except if

$$G^\varepsilon(x,t) = 0. \tag{9.57}$$

The latter equation determines the position of the moving reaction front, and $G(x,t) = \lim_{\varepsilon \to 0} G^\varepsilon(x,t)$ is considered as the principal Hamiltonian function. The Hamiltonian approach can be applied to calculate the velocity of the propagating front. The partial derivatives of $G(x,t)$ with respect to time and the coordinate have the physical meaning of the Hamiltonian and the momentum, see Eq. (8.54),

$$\frac{\partial G(x,t)}{\partial t} = -H, \qquad \frac{\partial G(x,t)}{\partial x} = p. \tag{9.58}$$

The ansatz (9.55) for the pdf inside the bulk is substituted into Eq. (9.53), where we also adopt the rescaling $x \to x/\varepsilon$ and $t \to t/\varepsilon$. We start with the last term in Eq. (9.53), the reaction term. First we take into account the following change in the upper bound of the integral,

$$\int_{-\infty}^{t} dt' \to \lim_{\varepsilon \to 0} \int_{-\infty}^{t/\varepsilon} dt' = \int_{-\infty}^{\infty} dt'. \tag{9.59}$$

Then we carry out the following expansion for the marginal pdf $P_1^\varepsilon(x, t/\varepsilon - t') \equiv P_1^\varepsilon(t - \varepsilon t')$:

$$P_1^\varepsilon(t - \varepsilon t') = \exp\left[-\frac{1}{\varepsilon}G^\varepsilon(t - \varepsilon t')\right] \approx \exp\left[-\frac{1}{\varepsilon}G^\varepsilon(t) + t'\partial_t G^\varepsilon(t)\right]. \tag{9.60}$$

Therefore the nonlinear reaction term reads

$$\text{NLRT} = \frac{1}{(2\pi i)^2} e^{-2G^\varepsilon/\varepsilon} \cdot \int_{-\infty}^{\infty} dt' dt''$$

$$\times \int_{\sigma-i\infty}^{\sigma+i\infty} \hat{\mathcal{R}}(s', s'') e^{-(H-s')t'} e^{-(H-s'')t''} ds' ds'', \quad (9.61)$$

where we used the first equation in (9.58). Integration with respect to s' and s'' is performed with some care, since $\hat{\mathcal{R}}$ is singular at $s' = \pm s''$. The main result is that $\hat{\mathcal{R}}$ is finite and that $\text{NLRT} \sim e^{-2G^\varepsilon/\varepsilon}$, which in the limit $\varepsilon \to 0$ is of the order of $o\left(e^{-G^\varepsilon/\varepsilon}\right)$. Therefore it does not contribute to the final result, since the rest of the equation is of the order of $O\left(e^{-G^\varepsilon/\varepsilon}\right)$.

Now we consider the kinetic term and carry out the integration with repsect to t with the kernel $\mathcal{R}_1$,

$$\int_0^{t/\varepsilon} \mathcal{R}_1(t') e^{-G^\varepsilon(t-\varepsilon t', x)/\varepsilon} dt'$$

$$= e^{-G^\varepsilon(t)/\varepsilon} \int_0^{\infty} \mathcal{R}_1(t') e^{-Ht'} dt' = e^{-G^\varepsilon(t)/\varepsilon} \hat{\mathcal{R}}_1(H), \quad (9.62)$$

where $\hat{\mathcal{R}}_1(H)$ is defined in Eq. (9.46).

Finally, differentiating in the limit $\varepsilon \to 0$, where $\text{NLRT} = 0$, and taking into account that the Hamiltonian H and the momentum p in Eq. (9.58) do not depend explicitly on x and t, which results in the absence of mixed derivatives, we find that the kinetic equation (9.53) becomes a Hamilton–Jacobi equation:

$$-\partial_t G = aD\left(\partial_x G\right)^2 \tilde{\mathcal{R}}_1(H) - C. \quad (9.63)$$

Here $G \equiv G(x, t) = \lim_{\varepsilon \to 0} G^\varepsilon$ is the action, or the principal Hamiltonian function,

$$G(x, t) = \int_0^t [p(\tau)\dot{x}(\tau) - H(p(\tau), x(\tau))]d\tau. \quad (9.64)$$

The velocity v of the propagating front is determined by the condition $G(x, t) = 0$. Together with the Hamilton equations, this yields

$$v = \dot{x} = \frac{\partial H}{\partial p}, \quad v = \frac{H}{p}. \quad (9.65)$$

The combination of these two equations can be replaced[3] by

$$v = \min_{H>0} \frac{H}{p(H)} = \min_{p>0} \frac{H(p)}{p}. \quad (9.66)$$

---

[3] Eqs. (9.65) follows from the derivation $\partial_p\left(H/p\right) = (\partial_p H)/p - H/p^2 = 0$.

To proceed, we take into account that $\partial_t G = -H$ and $\partial_x G = p$. We substitute these results into Eq. (9.63), and taking into account Eq. (9.46) for the kernel $\hat{\mathcal{R}}_1(H)$, we obtain Eq. (9.63) in the form

$$H = D_{\frac{1}{2}} p^2 \sqrt{H} \left[\tanh\left(\sqrt{H\,t_0} \right) \right]^{-1} - C. \tag{9.67}$$

The minimum condition, Eq. (9.66), for this equation yields

$$(3H + 5C) \cdot \sinh\left(2\sqrt{H\,t_0} \right) = 2(H + C)\sqrt{H\,t_0}. \tag{9.68}$$

Numerical solutions of Eq. (9.68) for H vs t_0 are presented in Fig. 9.2. These solutions determine the overall velocity v of the moving reaction front for different realizations of the experimental setup, reflected by the parameter $t_0 = h^2 b/ad$. The result of numerical calculations of the reaction front velocity is depicted in Fig. 9.3, where t_0 is the scaling parameter. It determines the dependence of the reaction front velocity v on the geometry of the experimental setups.

The solutions depend on the reaction rate C. For $C > 0$, no positive solutions exist for H, and the solution of Eq. (9.66) is $H = 0$ for all values of the scaling parameter t_0. For $C < 0$, there exist non-trivial solutions of Eqs. (9.66) and (9.68). We define two regimes with $t_0 > 1$ and $t_0 < 1$. The first one is the subdiffusive regime, where v is a constant function of t_0. In the second, the diffusive regime, the front velocity is a decreasing function of t_0, as it is shown in the inset of Fig. 9.3. The fitting curve, marked by stars, in the inset is $v(t_0) \sim 1/t_0^{1/4}$.

Analytical solutions of Eqs. (9.66) and (9.68) for the velocity can be obtained for limiting cases of t_0. As follows from the numerical solutions in Fig. 9.2, the physical values of the energy H are restricted to an interval of the order of $H \sim 1$. In this case, the asymptotic solutions of Eqs. (9.66) and (9.68) are determined by the limiting values of the scaling parameter t_0. For the analytical estimate, we consider two regimes with $t_0 \gg 1$ and $t_0 \ll 1$. The first one is the subdiffusive regime, where v is independent of t_0. In the second, the diffusive regime, the front velocity is a decreasing function of t_0, as shown in the inset of Fig. 9.3. In what follows we obtain analytical expressions for the velocities and the scaling $v = v(t_0)$.

9.5.1   Subdiffusive regime: $t_0 \gg 1$

In the subdiffusive regime we obtain

$$\tanh\left(\sqrt{H t_0} \right) \approx 1 \quad \text{and} \quad 2(H + C)\sqrt{H\,t_0}/\sinh\left(2\sqrt{H\,t_0} \right) \approx 0, \tag{9.69}$$

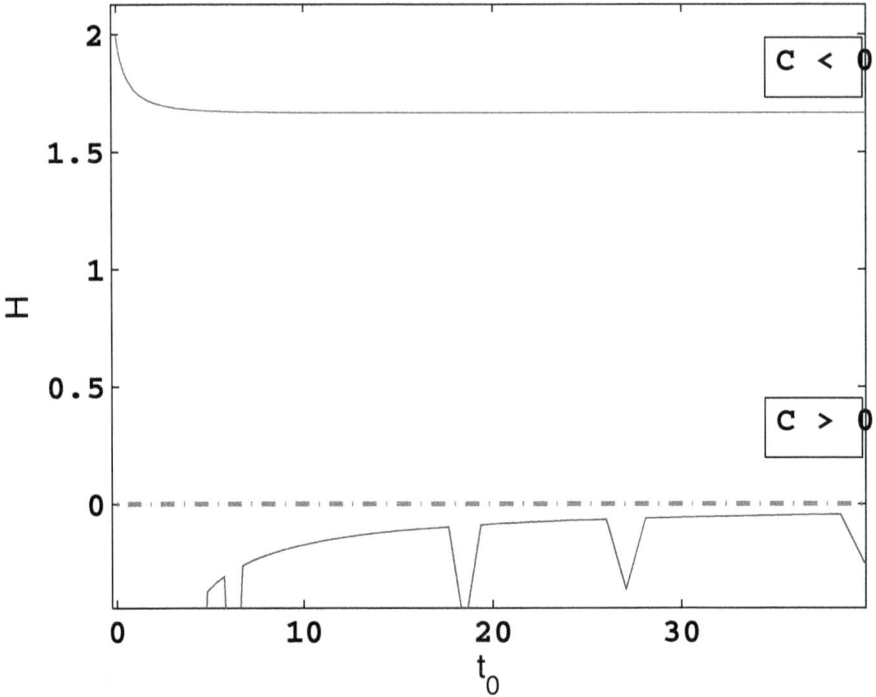

Fig. 9.2 Numerical solutions of Eq. (9.68) for the Hamiltonian H vs the transient time t_0 for $C = -1$ (upper plot) and $C = 1$ (lower plot). The transient time t_0 plays the role of a scaling parameter, which reflects various realizations of the comb geometry of the experimental setup, shown in Fig. 9.1. Reprinted from Chaos, Solitons & Fractals, 92, A. Iomin, V. Zaburdaev, and T. Pfohl, Reaction front propagation of actin polymerization in a comb-reaction system, 115–122 (2016) with permission from Elsevier.

which simplifies Eq. (9.68),

$$(3H + 5C) = 0. \tag{9.70}$$

This equation has no solution for $C > 0$. Therefore, for $C > 0$ and $H \geq 0$, the only solution for the front velocity is

$$v = \frac{H}{p(H)} = 0. \tag{9.71}$$

This means front propagation failure, as expected for the case where the initial concentration of the inducer is smaller than that of actin. The result is in complete agreement with numerical calculations. For $C < 0$, i.e., $P_0 >$

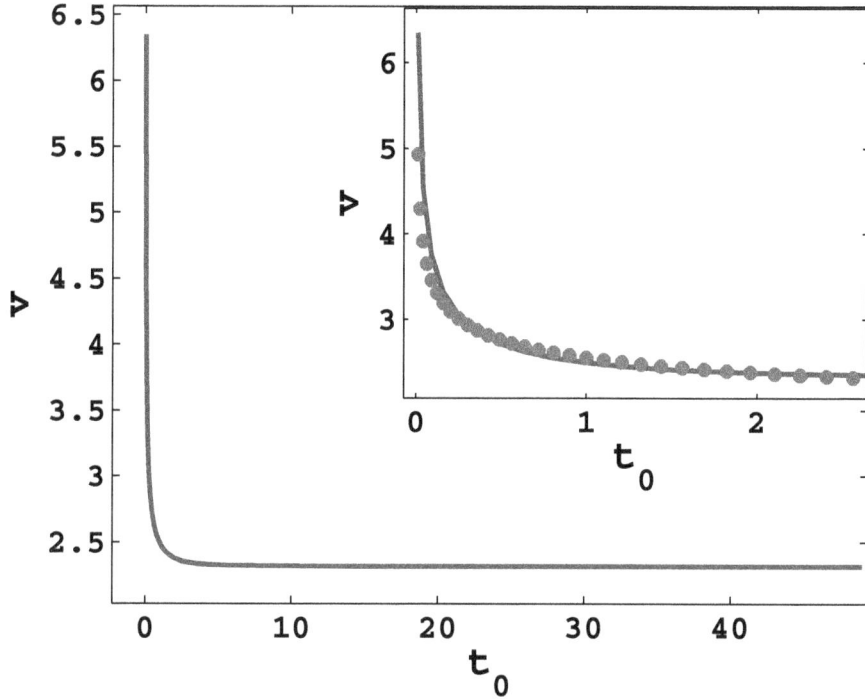

Fig. 9.3 The propagation velocity v of the reaction front for different realizations of the experimental setup t_0 for $C = -1$ and $D_{1/2} = 1$. The inset describes the result for the small values of t_0, solid line. The fitting plot (*) corresponds to $t_0^{-1/4}$ curve. Reprinted from Chaos, Solitons & Fractals, 92, A. Iomin, V. Zaburdaev, and T. Pfohl, Reaction front propagation of actin polymerization in a comb-reaction system, 115–122 (2016) with permission from Elsevier.

N, there exists a solution, $H = 5|C|/3$, which yields a nonzero propagation velocity of the reaction front in the subdiffusive regime,

$$v = \frac{H^{5/4}\sqrt{D_{\frac{1}{2}}}}{\sqrt{H + C}} = \left(\frac{5}{3}\right)^{\frac{5}{4}}\left(\frac{3}{2}\right)^{\frac{1}{2}}\left(D_{\frac{1}{2}}^2|C|^3\right)^{\frac{1}{4}} = \left(\frac{5}{3}\right)^{\frac{5}{4}}\left(\frac{3}{2}\right)^{\frac{1}{2}}\left(\frac{D^2ab|C|^3}{2d}\right)^{\frac{1}{4}}.$$

$$(9.72)$$

As expected, this result depends on the diffusivity of the reagents and the geometrical parameters a and b. However, it is independent of the finger length, which makes it possible to arrive at the correct comb limit $h \to \infty$, when only subdiffusion takes place.

9.5.2   *Diffusive regime:* $t_0 \ll 1$

In the opposite case of the diffusive regime, when $t_0 \ll 1$, we have

$$\tanh\left(\sqrt{H/d_\rho}\right) \approx \sinh\left(\sqrt{H/d_\rho}\right) \approx h\sqrt{H/d_\rho}. \tag{9.73}$$

For $C < 0$ ($P_0 > N$), Eq. (9.68) implies that $H = 2|C|$. Substituting this result into $v = H/p(H)$, we obtain

$$v = 2\sqrt{|C|D_{\frac{1}{2}}} \cdot t_0^{-\frac{1}{4}}. \tag{9.74}$$

For $C > 0$, no solution of Eq. (9.68) exists, and the front velocity vanishes, $v = 0$.

The same results can be obtained from Eq. (9.63), which in this limit reads

$$H = \bar{D}p^2 - C, \tag{9.75}$$

where $\bar{D}$ is defined in Eq. (9.37) and $p = \sqrt{(H+C)/\bar{D}}$. This immediately yields

$$v = \min_{H>0}\left[\frac{H\bar{D}^{\frac{1}{2}}}{\sqrt{(H+C)}}\right] = 0 \tag{9.76}$$

for $H = 0$, as expected for $C > 0$.

The situation changes dramatically for $C < 0$. In this case, the equation

$$\frac{\partial H}{\partial p} = \frac{H}{p} \tag{9.77}$$

yields

$$2\bar{D}p = \bar{D}p + \frac{|C|}{p}, \tag{9.78}$$

and the solution is $p = \sqrt{|C|/\bar{D}}$. After substitution into the velocity equation we find

$$v = \min_{p>0}\frac{H(p)}{p} = \frac{\bar{D}p^2 + |C|}{p} = 2\sqrt{\bar{D}|C|}. \tag{9.79}$$

This is the well known FKPP result for the reaction front velocity with the scaled diffusivity $\bar{D}$. Taking into account that $D_{\frac{1}{2}} = \bar{D}\sqrt{t_0}$, we obtain Eq. (9.74). As shown above, in this regime the front velocity scales like $v \sim t_0^{-1/4} = \left(da/h^2b\right)^{1/4}$. This analytical result agrees with the numerical one. Both results are depicted in the inset of Fig. 9.3.

9.5.3 *Discussion*

The evaluation of the propagation velocity of the reaction front in the framework of the hyperbolic scaling is in complete agreement with the exact solution of the extinction dynamics. Indeed, as follows from solutions (9.34) and (9.38) for $C > 0$, when the actin concentration is larger than that of the inducer, the velocity of the reaction front is zero. These results coincide completely with the numerical and analytical results obtained by the hyperbolic scaling in the framework of the Hamilton–Jacobi equation, when $v = 0$. For $C < 0$, the nonlinear term becomes important. As follows from the solution (9.38), the velocity of the reaction front, obtained from the condition $|C|t - x^2/4\bar{D}t = 0$, yields the result of Eq. (9.79) $v = 2\sqrt{\bar{D}|C|}$. For FKPP dynamics, the linearization of the nonlinear reaction-transport equation leads to a good approximation for the propagation velocity of the reaction front. For subdiffusive dynamics, the situation is more complex for $C < 0$, since the linearization approach is not valid. In this case, the hyperbolic scaling in the framework of the Hamilton–Jacobi equation is the reliable approach to the problem.

PART 4

Extensions of the Comb Model

Chapter 10

Lévy processes on a generalized fractal comb

10.1 Introduction

The dimensionality of the medium, expressed in the form of a fractal dimension ν plays an important role in the transport characteristics of random processes. So far, we have considered the comb model in two dimensional space (except Chapter 7), where the fingers are continuously and homogeneously distributed along the x-axis. The "comb geometry" is due to the inhomogeneous properties of the transport coefficient along the x-direction, namely $D_{xx} = D_x \delta(y)$, which ensures the absence of transport for $y \neq 0$. What happens to the dynamics, if the fingers have an inhomogeneous, e.g., a fractal distribution, as shown in Fig. 10.1? To understand the influence of the fractal structure with fractal dimension ν, we consider first the simpler case of a fractal grid comb [157].

10.2 Fractional diffusion on a fractal grid comb

We consider a generalization of Eq. (1.1), where diffusion along x-direction occurs on many backbones, located at positions $y = l_j$, which belong to a fractal set $\mathcal{S}_\nu$ with fractal dimension $0 < \nu < 1$. The transport equation for such a structure reads

$$\frac{\partial}{\partial t} P(x, y, t) = D_x \sum_{l_j \in \mathcal{S}_\nu} \delta(y - l_j) \frac{\partial^2}{\partial x^2} P(x, y, t) + D_y \frac{\partial^2}{\partial y^2} P(x, y, t). \quad (10.1)$$

We employ the same initial and boundary conditions as for Eq. (1.1) and also note that D_x is a generalized diffusion coefficient with physical dimension $[D_x] = \text{m}^{3-\nu}/\text{s}$, which absorbs the dimension of the fractal volume or measure μ. The procedure of solving Eq. (10.1) is the same as for Eq. (1.7).

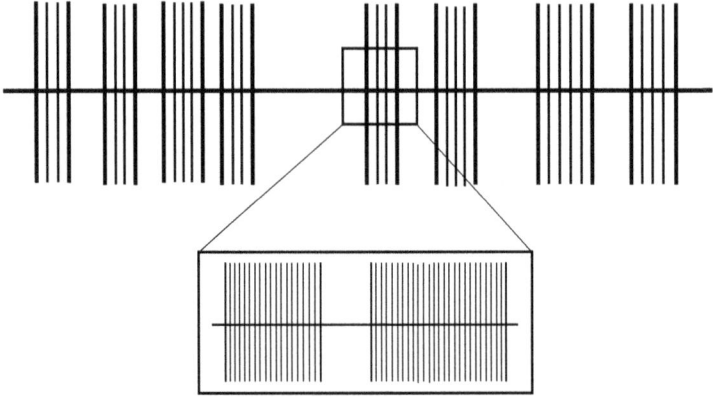

Fig. 10.1   Fourth generation of the random one-third Cantor set. This fractal comb is a form of a middle-third Cantor set construction, where each segment is randomly divided in three parts. The second slice is the first generation of the smallest part of the third generation of the Cantor set, shown in the upper slice. Republished with permission of IOP Publishing, LTD, from J. Phys. A: Math. Theor., T. Sandev, A. Iomin, and V. Méndez, 49(35), 355001 (2016); permission conveyed through Copyright Clearance Center, Inc.

After the Laplace–Fourier transformation of Eq. (10.1), we look for the solution in the form

$$\hat{\tilde{P}}(k,y,s) = \hat{\tilde{f}}(k,s)\exp\left(-\sqrt{s/D_y}|y|\right),\qquad(10.2)$$

where $\hat{\tilde{P}}(k, y = l_j, s) = e^{-\sqrt{s/D_y}|l_j|}\hat{\tilde{f}}(k,s)$ describes diffusion along the backbone at $y = l_j$ of the fractal structure $\mathcal{S}_\nu$ in Laplace–Fourier space. We obtain from Eqs. (10.1) and (10.2) for the marginal pdf

$$s^{\frac{1}{2}}\hat{\tilde{P}}_1(k,s) = -k^2 D_{\frac{1}{2}}\sum_{l_j\in\mathcal{S}_\nu} e^{-\sqrt{s/D_y}|l_j|}\hat{\tilde{P}}_1(k,s) + s^{-\frac{1}{2}},\qquad(10.3)$$

where $D_{\frac{1}{2}} = \sqrt{D_x^2/4D_y}$, see Eq. (9.29).

It remains to calculate the sum $\sum_{l_j\in\mathcal{S}_\nu} e^{-\sqrt{s/D_y}|l_j|}$. Note that fractal sets, like a Cantor set, are uncountable. Therefore, the last expression is purely formal and its mathematical realization corresponds to the integration with the fractal measure $\mu_\nu \sim l^\nu$, such that $\sum_{l_j\in\mathcal{S}_\nu} \delta(l-l_j) = \frac{1}{\Gamma(\nu)}l^{\nu-1}$ is the fractal density [42], and $d\mu_\nu = \frac{1}{\Gamma(\nu)}l^{\nu-1}dl$, which yields

$$\frac{2}{\Gamma(\nu)}\int_0^\infty dl\, l^{\nu-1}e^{-\sqrt{s/D_y}l} = \left(\frac{D_y}{s}\right)^{\nu/2}.\qquad(10.4)$$

Here we expand the fractal set, embedded in the segment $[-L, L]$ on the infinite line, keeping the fractal dimension ν the same. We explain this procedure in Sec. 10.4.1.

Eventually, we obtain the FFPE with the Caputo fractional time derivative $\partial_t^\alpha \equiv \frac{\partial^\alpha}{\partial t^\alpha}$ of the order of $\alpha = (1+\nu)/2$ and the diffusion coefficient $D_{\bar\alpha} \equiv D_{\frac{1-\nu}{2}} = D_x / \sqrt{D_y^{1-\nu}}$:

$$\partial_t^\alpha P_1(x,t) = D_{\bar\alpha} \frac{\partial^2 P(x,t)}{\partial x^2}. \tag{10.5}$$

The MSD reads

$$\left\langle x^2(t) \right\rangle = \mathcal{L}^{-1}\left[-\frac{\partial^2}{\partial k^2} \hat{\tilde{P}}_1(k,s) \bigg|_{k=0} \right] = \frac{D_{\bar\alpha} t^\alpha}{\Gamma(1+\alpha)} = \frac{D_x}{D_y^{\frac{1-\nu}{2}}} \frac{t^{\frac{1+\nu}{2}}}{\Gamma\left(1 + \frac{1+\nu}{2}\right)}, \tag{10.6}$$

which describes anomalous diffusion with the transport exponent α, $\frac{1}{2} < \alpha < 1$. The fractal set $\mathcal{S}_\nu$ of the infinite number of backbones increases the transport exponent from $1/2$ to $\frac{1+\nu}{2}$. For $\nu = 1$, the MSD becomes $\left\langle x^2(t) \right\rangle \simeq t$, which is consistent with two-dimensional normal diffusion, and for $\nu = 0$ we are back to the cases of either a single backbone or a finite number of backbones. Note that the fractal dimension of any finite number of discrete points is $\nu = 0$.

10.3 Auxiliary: Fractals and Fractal Dimension

The concept of a fractal was introduced by mathematicians, such as Cantor and Peano, at the end of the 19th century. The relation of fractals to natural phenomena was emphasized by Mandelbrot [158], who also coined the term "fractal" in 1975. This idea was first expressed by Kolmogorov in Ref. [159] for fractional Brownian motion, and later rediscovered by Mandelbrot and Van Ness, who explored this phenomenon in greater detail in Ref. [160]. A fractal is a set which has a fractal dimension. To make this definition precise, it is necessary to define a concept of dimension applicable to a large class of sets. Mathematically, a fractal can be defined as a set of points in a d-dimensional metric space for which it is impossible to determine any conventional measure with integer dimension, i.e., length, area, or volume. The rigorous mathematical concept of the fractal dimension, the so-called Hausdorff–Besicovich dimension $\dim_H \mathcal{S}$, relates to the fractional Hausdorff measure [161] $\mu_\nu(\mathcal{S})$ of a set $\mathcal{S}$ such that

$$\dim_H \mathcal{S} = \inf\{\nu : \ \mu_\nu(\mathcal{S}) = 0\} = \sup\{\nu : \mu_\nu(\mathcal{S}) = \infty\}, \tag{10.7}$$

which yields the definition

$$\mu_\nu(\mathcal{S}) = \begin{cases} \infty & \text{if } \nu > \dim_H S, \\ 0 & \text{if } \nu < \dim_H S. \end{cases} \tag{10.8}$$

The mathematical concept of the Hausdorff–Besicovich dimension is not really useful for most physical applications. In physics, fractals can be characterized not only by the Hausdorff–Besicovich dimension but by a number of other dimensions that are easy to find in experiments, real or numerical, and that provide a versatile description of the object's properties. One possible simplified dimension, proposed by Kolmogorov, is the *capacity* or *fractal dimension* related to box counting.[1] Let $\mathcal{S} \subset R^d$ be a compact set. For every $\epsilon > 0$ denote by $N(\epsilon)$ the minimum number of d-dimensional hypercubes, cubes of linear size ϵ, needed to cover $\mathcal{S}$. The fractal dimension of $\mathcal{S}$ is defined by the formula

$$\nu_f = \dim \mathcal{S} = \lim_{\epsilon \to 0} \frac{\log N(\epsilon)}{\log(1/\epsilon)}, \tag{10.9}$$

if the limit exists. This coincides with the *scaling dimension*, which is defined when $\mathcal{S}$ can be subdivided into P replicas of itself, each of which is similar to $\mathcal{S}$ and with linear magnification by a factor Q becomes congruent to $\mathcal{S}$. The scaling dimension is given by

$$\nu_s = \dim \mathcal{S} = \frac{\log P}{\log Q}. \tag{10.10}$$

Another important definition of the fractal set dimension, which is important for random walks in the large scale asymptotics, is the mass dimension. Assigning a unit density to $\mathcal{S}$, we define the mass of the set of radius l as $m(l) \sim l^\nu$. We can also define it as a measure or volume: $\mu(l) \sim l^\nu$. Note that $\nu = \nu_f = \nu_s$, and we call it the fractal dimension of a fractal set $\mathcal{S}_\nu$.

10.4 Fractal comb

Another natural phenomenological extension of the comb equations (1.1) and (8.24) is the generalization of both time processes by introducing memory kernels $\gamma(t)$ and $\eta(t)$ and considering a fractal spatial geometry, i.e., a power-law density of the fingers described by a kernel $\rho(x)$. This modification of the comb model (1.1) corresponds to a so-called fractal comb model,

---

[1]There is vast literature on the subject. We cite only few references where this subject is well discussed for random walks and chaotic dynamics [158, 162, 163, 80].

$$\int_0^t dt'\, \gamma(t-t')\frac{\partial}{\partial t'}P(x,y,t') = D_x\delta(y)\int_0^t dt'\, \eta(t-t')\frac{\partial^2}{\partial x^2}P(x,y,t')$$

$$+ D_y\frac{\partial^2}{\partial y^2}\int_{-\infty}^{\infty} dx'\, \rho(x-x')P(x',y,t).$$

$$(10.11)$$

Here, the memory kernels $\gamma(t)$ and $\eta(t)$ are in general decreasing functions that approach zero in the long time limit.[2] The physical dimensions of the diffusion coefficients $D_x\delta(y)$ and D_y depend now on the form of the memory kernels $\gamma(t)$ and $\eta(t)$. The memory kernels $\gamma(t)$ and $\eta(t)$, and the kernel[3] $\rho(x)$ should be introduced in such a way that these functions do not change the physical meaning of the diffusion coefficients $D_x\delta(y)$ and D_y. Therefore it is convenient to write these functions in dimensionless form, by introducing the time scale τ and the coordinate scale l. This can be achieved in the same way as in Chapter 6 for Eq. (6.11): $\tau = D_x^2/D_y^3$ and $l = D_x/D_y$. This yields the corresponding change of the kernels $\gamma(t/\tau)$, $\eta(t/\tau)$, and $\rho(x/l)$, and leads to the rescaling of Eq. (10.11). For simplicity and to keep the diffusion coefficients D_x and D_y explicitly, we imply in the following that the diffusion coefficients automatically absorb these scale parameters. This rescaling depends on the functional form of $\gamma(t)$, $\eta(t)$, and $\rho(x)$. The function $\gamma(t)$ contributes to the memory effects in such a way that particles moving along the y-direction, i.e., along the fingers, can be trapped. This means that diffusion along the y-direction can be anomalous as well [16], see Chapter 8. We call the function $\eta(t)$ a generalized compensation kernel [16, 165], see Sec. 10.7 for an explanation of this terminology. The case $\gamma(t) = \eta(t) = \delta(t)$ yields the diffusion equation of the comb model (1.1).

The spatial fractal geometry is taken into account by the fractal dimension of the finger volume or mass, $|x|^\nu$, where $0 < \nu < 1$ is the fractional dimension, and an infinite number of fingers is continuously distributed by the power-law. This can be represented as a convolution integral between the non-local density of the fingers and the pdf $P(x,y,t)$ in the form $\int_{-\infty}^{\infty} dx'\, \rho(x-x')P(x',y,t)$ [166]. The important feature of this convolution integral relates to the scaling property of $\rho(x)$, which obeys the relation $\rho(x) = \frac{1}{a}\rho(bx)$, describing a fractal set with fractal dimension $\nu = \frac{\ln(a)}{\ln(b)}$ and log-periodicity with period $\ln(b)$. An example of a fractal comb is shown in Fig. 10.1. However, to consider anomalous diffusion on the fractal comb, we first need to construct this Cantor set on the infinite x-axis.

---

[2]More details on the form of the memory kernels can be found in Ref. [164].
[3]Note that the density of fingers is $\int dx\, \rho(x)$.

10.4.1   *Fractal set of fingers*

The construction of the fractal set $\mathcal{S}_\nu$ of the backbones, or the fingers, on the finite segment $[-L, L]$ is straightforward. However, construction of the Cantor set on the infinite axis requires some care. We propose the following algorithm of construction of a random middle-third Cantor set on the infinite x-axis. It is based on the fact that the cardinality of the finite segment $[0, 1]$ is the same as the cardinality of an infinite line.[4]

First, we represent the unit segment, which contains the fractal set, in the form of a circle by closing the end points. The end points are marked by O on the top of the circle in Fig. 10.2. The embedded fractal set is built by the standard procedure of a random one-third Cantor set construction: the circle is randomly divided into three parts by points th O, A_1 and B_1, and the central segment A_1B_1 is removed. From point O we draw rays passing through A_1 and B_1, which hit the horizontal line. Next, we divide each of the segments OA_1 and OB_1 into three parts by the points A_2 and A_3, and B_2 and B_3, respectively, removing the central segments A_2A_3 and B_2B_3. We repeat this procedure *ad infinitum*, cutting out the subintervals

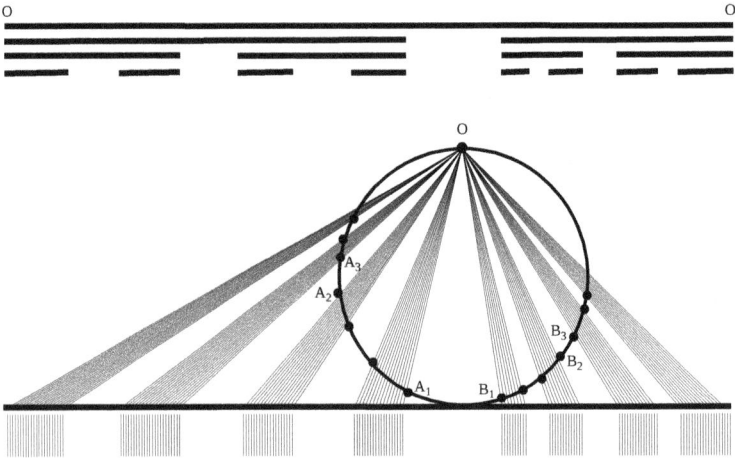

Fig. 10.2   Construction of a fractal grid of the fingers on an infinite axis. The top slice is the forth generation of the Cantor set construction. Reprinted figure with permission from T. Sandev, A. Iomin, and H. Kantz, Phys. Rev. E, 95, 052107 (2017). Copyright 2017 by the American Physical Society.

---

[4]In fact, there is a one-to-one correspondence between the Cantor set and all points in the unit segment [163], although the Lebesgue measures of the fractal set and the unit interval are 0 and 1, respectively.

and sending rays through the end points to construct the Cantor set on the infinite line with the same fractal dimension. Considering the fingers on the horizontal line, we construct a fractal set of fingers on the infinite x-axis. It is worth noting that the end points O of the unit segment, belonging to the circle fractal set, also belong to the fractal set of the x-axis since the top point O is mapped onto the horizontal line corresponding to $\pm\infty$. Therefore, this projection also ensures the existence of the boundary conditions at infinity.

10.5 Fractal structure of fingers and the Weierstrass function

A variety of fractal distributions of fingers, $\rho(x)$, has been studied in Ref. [165]. Here, we consider one example related to the Weierstrass function. We represent the last term in Eq. (10.11) in the form of the superposition

$$D_y \frac{\partial^2}{\partial y^2} \int_{-\infty}^{\infty} dx' \, \rho(x - x') P(x', y, t)$$

$$\Longrightarrow D_y \frac{\partial^2}{\partial y^2} \sum_{j=1}^{\infty} w_j \delta(x - l_j) P(x, y, t)$$

$$= D_y \frac{\partial^2}{\partial y^2} \sum_{j=1}^{\infty} w_j \mathcal{F}_k^{-1} \left[\int_{-\infty}^{\infty} dk' e^{i(k-k')l_j} \tilde{P}(k', y, t) \right], \qquad (10.12)$$

and finally as a convolution in Fourier space. This form of the diffusion term that describes transport along the y-direction exactly expresses the fact that diffusion along the fingers occurs on the infinite number of fingers located at $x = l_j$, $j = 1, 2, ..., 0 \le l_1 < l_2 < \cdots < l_N < \cdots$, i.e., positions on the x-axis that belong to the fractal set S_ν with fractal dimension $0 < \nu < 1$. The w_j are structural constants such that $\sum_{j=1}^{\infty} w_j = 1$.

To obtain the Weierstrass function we follow the procedure given in Ref. [167]. We display fingers on the segment $[0, L]$, such that $l_j = L/l^j$, exploiting the one-to-one correspondence between the fractal set embedded in the segment and the fractal set obtained by mapping it onto the infinite x axis. Choosing the structural constants as follows, $w_j = \frac{l-b}{b} \left(\frac{b}{l}\right)^j$, where $l, b > 1$, $l - b \ll b$ (l and b are dimensionless scale parameters), we find

$$\sum_{j=1}^{\infty} w_j = \frac{l-b}{l} \sum_{j=0}^{\infty} \left(\frac{b}{l}\right)^j = 1. \qquad (10.13)$$

From Eqs. (10.12) and (10.13) we have

$$\Psi(z) = \frac{l-b}{l} \sum_{j=1}^{\infty} \left(\frac{b}{l}\right)^j \exp\left(i\frac{z}{lj}\right), \qquad (10.14)$$

where $z = (k - k')L$ and $l_1 = L = 1$. After some simple algebra we obtain

$$\Psi(z/l) = \frac{l}{b}\Psi(z) - \frac{l-b}{b}\exp\left(i\frac{z}{l}\right). \qquad (10.15)$$

Neglecting the last term ($l - b \ll b$), we find the following scaling relation,

$$\Psi(z/l) \simeq \frac{l}{b}\Psi(z). \qquad (10.16)$$

This means that $\Psi(z) \sim (z^{1-\bar{\nu}})^{-1}$, where $\bar{\nu} = \log b / \log l$, $0 < \bar{\nu} < 1$, is the fractal dimension.

Consequently, the convolution representation of the finger diffusion term is the convolution with the Weierstrass functions, omitting the negligibly small term in Eq. (10.15), which eventually is represented in the form of the Riesz fractional integral in reciprocal Fourier space,

$$D_y \frac{\partial^2}{\partial y^2} \frac{1}{2\pi} \int_{-\infty}^{\infty} d\kappa \Psi(k - \kappa) \tilde{P}(\kappa, y, t) = D_y \frac{1}{2\pi} \frac{\partial^2}{\partial y^2} \int_{-\infty}^{\infty} d\kappa \frac{\tilde{P}(\kappa, y, t)}{|k - \kappa|^{1-\bar{\nu}}}. \qquad (10.17)$$

10.5.1 *Fractional diffusion solution for the fractal comb*

Applying the inverse Fourier transformation with respect to k in Eq. (10.12) and changing the order of integration, we obtain

$$D_y \frac{\partial^2}{\partial y^2} \mathcal{F}_k^{-1}\left[\int_{-\infty}^{\infty} d\kappa \frac{\tilde{P}(\kappa, y, t)}{|k - \kappa|^{1-\bar{\nu}}}\right] = D_y C_\nu |x|^{-\bar{\nu}} \frac{\partial^2}{\partial y^2} P(x, y, t), \qquad (10.18)$$

where $C_{\bar{\nu}} = \Gamma(\bar{\nu}) \cos \frac{\bar{\nu}\pi}{2}$ is a normalization constant. The convolution with the Weierstrass function is normalized with $C_{\bar{\nu}}$ to obtain the correct form of the Riesz fractional integration in Eq. (10.17). In the following analysis, this multiplier disappears as the result of the normalization procedure, and Eq. (10.11) becomes

$$\int_0^t dt' \, \gamma(t - t') \frac{\partial}{\partial t'} P(x, y, t') = D_x \delta(y) \int_0^t dt' \, \eta(t - t') \frac{\partial^2}{\partial x^2} P(x, y, t')$$

$$+ D_y |x|^{-\bar{\nu}} \frac{\partial^2}{\partial y^2} P(x, y, t). \qquad (10.19)$$

Note that the continuous comb model corresponds to the limit with $\bar{\nu} = 0$. In this sense $\bar{\nu}$ is dual to ν with the relation $\bar{\nu} = 1 - \nu$. Therefore we consider a fractal comb with the fractal density of the fingers scaling like $|x|^{\nu-1}$, which corresponds to the fractal set of the fingers S_ν with fractal dimension ν.

Performing the Laplace transformation of Eq. (10.19), we obtain

$$\hat{\gamma}(s)\left[s\hat{P}(x,y,s) - \delta(x)\delta(y)\right]$$

$$= D_x\delta(y)\hat{\eta}(s)\frac{\partial^2}{\partial x^2}\hat{P}(x,y,s) + D_yC_{\bar{\nu}}|x|^{-\bar{\nu}}\frac{\partial^2}{\partial y^2}\hat{P}(x,y,s). \qquad (10.20)$$

The next, standard, step is to look for the solution of Eq. (10.20) in the form

$$\hat{P}(x,y,s) = \exp\left(-\sqrt{\frac{s\hat{g}(x,s)}{D_y}}|y|\right)f(x,s), \qquad (10.21)$$

where $\hat{g}(x,s)$ is obtained from the condition that the second derivative of the exponential in Eq. (10.21) compensates the first term on the lhs of Eq. (10.20). This yields

$$\hat{g}(x,s) = \hat{\gamma}(s)|x|^{\bar{\nu}}, \qquad (10.22)$$

and the solution $\hat{P}(x,y,s)$ becomes

$$\hat{P}(x,y,s) = \exp\left(-\sqrt{\frac{s\hat{\gamma}(s)}{D_y}}|x|^{\bar{\nu}/2}|y|\right)f(x,s). \qquad (10.23)$$

From here we find the relation between the marginal pdf $\hat{P}_1(x,s)$ and the pdf at $y = 0$, $\hat{P}(x,y=0,s) = f(x,s)$,

$$\hat{P}_1(x,s) = \int_{-\infty}^{\infty} dy\, \hat{P}(x,y,s) = 2\sqrt{\frac{D_y}{s\hat{g}(x,s)}}f(x,s). \qquad (10.24)$$

Integrating Eq. (10.20) with respect to y and taking into account Eq. (10.24), we obtain the boundary value problem for the marginal pdf $\hat{P}_1(x,s)$. However, it is more convenient to consider this problem for $f(x,s)$, which immediately leads to the equation for the Green's function $f(x,s)$ with zero boundary conditions at infinity,

$$2\sqrt{\frac{D_ys}{\hat{\gamma}(s)}}|x|^{-\bar{\nu}/2}f(x,s) - D_x\frac{\hat{\eta}(s)}{\hat{\gamma}(s)}\frac{\partial^2}{\partial x^2}f(x,s) = \delta(x). \qquad (10.25)$$

This equation is symmetric with respect to the reflection $x \to -x$. Therefore we look for the symmetrical solution $f(x,s) = C_{\bar{\nu}}G(|x|,s) \equiv C_{\bar{\nu}}G(r,s)$, where $r = |x|$ and $C_{\bar{\nu}}$ is a constant, depending on the parameter s. Substituting this form into Eq. (10.25), we obtain for the second derivative

$$\frac{d^2G(r,s)}{dx^2} = \frac{d^2G(r,s)}{dr^2} \cdot \left(\frac{d|x|}{dx}\right)^2 + \frac{dG(r,s)}{dr} \cdot \left(\frac{d^2|x|}{dx^2}\right)$$

$$= \frac{d^2G(r,s)}{dr^2} + 2\delta(x)\frac{dG(r,s)}{dr}. \tag{10.26}$$

Taking this result into account, we arrive at the standard procedure by considering the homogeneous part of Eq. (10.25),

$$D_{\frac{1}{2}}^{-1}\frac{\sqrt{s\hat{\gamma}(s)}}{\hat{\eta}(s)}r^{-\bar{\nu}/2}G(r,s) - \frac{\partial^2}{\partial r^2}G(r,s) = 0, \tag{10.27}$$

and the shift condition at $r = 0$ related to the $\delta(x)$ function,

$$-2D_x C_{\bar{\nu}} \left.\frac{dG(r,s)}{dr}\right|_{r=0} = 1. \tag{10.28}$$

Here we also introduced the generalized kinetic coefficient

$$D_{\frac{1}{2}} = \frac{D_x}{2\left[D_y\right]^{1/2}}. \tag{10.29}$$

To be specific,[5] we consider first the special case with $\hat{\gamma}(s) = \hat{\eta}(s) = 1$. Then Eq. (10.27) reads

$$D_{\frac{1}{2}}^{-1}s^{1/2}r^{-\bar{\nu}/2}G(r,s) = \frac{\partial^2 G(r,s)}{\partial r^2}, \tag{10.30}$$

which has the form of the Bessel differential equation [125].

Let us consider the Bessel-form equation

$$G''(r) - c^2 r^{2\zeta-2}G(r) = 0, \tag{10.31}$$

where $c^2 = D_{\nu}^{-1}s^{1/2}$ and $\zeta = \frac{4-\bar{\nu}}{4}$. Its solution is given in terms of Bessel functions [125],

$$G(r) = \sqrt{r}Z_{\frac{1}{2\zeta}}\left(i\frac{c}{\zeta}r^{\zeta}\right), \tag{10.32}$$

where $Z_{\frac{1}{2\zeta}}(r) = C_1 J_{\frac{1}{2\zeta}}(r) + C_2 N_{\frac{1}{2\zeta}}(r)$ is a combination of the Bessel function of the first kind $J_{\frac{1}{2\zeta}}(r)$ and the Bessel function of the second kind (Neumann function) $N_{\frac{1}{2\zeta}}(r)$.

Therefore the solution of Eq. (10.30) reads

$$\hat{G}(|x|,s) = \sqrt{|x|}Z_{\frac{2}{4-\bar{\nu}}}\left(i\frac{4}{4-\bar{\nu}}D_{\nu}^{-1/2}s^{1/4}|x|^{\frac{4-\bar{\nu}}{4}}\right). \tag{10.33}$$

---

[5]In this section we focus on the fractal property of the fingers and consider the memory kernels to be delta functions: $\gamma(t-t') = \eta(t-t') = \delta(t-t')$.

Due to the zero boundary conditions, Green's function (10.33) is given by the modified Bessel function [46] $K_{\frac{2}{4-\bar{\nu}}}(z)$, which can be expressed in terms of the Fox H-function as well, see relation (2.45),

$$
\begin{aligned}
G(|x|, s) &= \sqrt{|x|} K_{\frac{2}{4-\bar{\nu}}}\left(\frac{4}{4-\bar{\nu}} D_{\frac{1}{2}}^{-1/2} s^{1/4} |x|^{\frac{4-\bar{\nu}}{4}}\right) \\
&= \frac{\sqrt{|x|}}{2} H_{0,2}^{2,0}\left[\frac{4}{(4-\bar{\nu})^2} D_{\frac{1}{2}}^{-1} s^{1/2} |x|^{\frac{4-\bar{\nu}}{2}} \, \middle| \, \left(\frac{1}{4-\bar{\nu}}, 1\right), \left(-\frac{1}{4-\bar{\nu}}, 1\right)\right].
\end{aligned}
$$
(10.34)

To find the coefficient $\mathcal{C}_{\bar{\nu}}(s)$, we use the series representation of the Bessel function $K_\nu(z)$ for $z \to 0$ and $\nu \notin \mathbb{Z}$,

$$
\begin{aligned}
K_\nu(z) &\simeq \frac{\Gamma(\nu)}{2}\left(\frac{z}{2}\right)^{-\nu}\left[1 + \frac{z^2}{4(1-\nu)} + \cdots\right] \\
&\quad + \frac{\Gamma(-\nu)}{2}\left(\frac{z}{2}\right)^{\nu}\left[1 + \frac{z^2}{4(\nu+1)} + \cdots\right].
\end{aligned}
$$
(10.35)

The nonzero contribution near $r = 0$ is

$$
G(r, s) \approx -r s^{\frac{1}{2(4-\bar{\nu})}} \frac{4 - \bar{\nu}}{4\left(\frac{2-\bar{\nu}}{4-\bar{\nu}}\right)}\left(\frac{4}{(4-\bar{\nu})^2} D_{\frac{1}{2}}^{-1}\right)^{-\frac{1}{4-\bar{\nu}}},
$$
(10.36)

where we used the relation for the gamma function

$$
\Gamma\left(-\frac{2}{4-\bar{\nu}}\right) = -\frac{4-\bar{\nu}}{2}\Gamma\left(\frac{2-\bar{\nu}}{4-\bar{\nu}}\right).
$$
(10.37)

Substituting the solution (10.34) into Eq. (10.25), and using relations (10.28) and (10.35), we find

$$
\mathcal{C}_{\bar{\nu}}(s) = \frac{2}{4-\bar{\nu}} \frac{1}{\Gamma\left(\frac{2-\bar{\nu}}{4-\bar{\nu}}\right) D_x}\left(\frac{4}{(4-\bar{\nu})^2} D_{\frac{1}{2}}^{-1}\right)^{-\frac{1}{4-\bar{\nu}}} s^{-\frac{1}{2(4-\bar{\nu})}},
$$
(10.38)

which yields the solution of Eq. (10.25). Taking into account the relation (10.24), we obtain the marginal pdf $\hat{P}_1(x, s)$,

$$
\begin{aligned}
\hat{P}_1(x, s) &= \frac{1}{4-\bar{\nu}} \frac{D_{\frac{1}{2}}^{-1}}{\Gamma\left(\frac{2-\bar{\nu}}{4-\bar{\nu}}\right)}\left(\frac{4 D_{\frac{1}{2}}^{-1}}{(4-\bar{\nu})^2}\right)^{-\frac{1}{4-\bar{\nu}}} s^{-\frac{5-\bar{\nu}}{2(4-\bar{\nu})}} |x|^{\frac{1-\bar{\nu}}{2}} \\
&\quad \times H_{0,2}^{2,0}\left[\frac{4 D_{\frac{1}{2}}^{-1}}{(4-\bar{\nu})^2} |x|^{\frac{4-\bar{\nu}}{2}} s^{1/2} \, \middle| \, \left(\frac{1}{4-\bar{\nu}}, 1\right), \left(-\frac{1}{4-\bar{\nu}}, 1\right)\right].
\end{aligned}
$$
(10.39)

We apply the inverse Laplace transformation formula (2.53) for the Fox H-function,

$$\mathcal{L}^{-1}\left[s^{-\rho}H_{p,q}^{m,n}\left[as^{\sigma}\left|\begin{array}{c}(a_p, A_p)\\(b_q, B_q)\end{array}\right.\right]\right] = t^{\rho-1}H_{p+1,q}^{m,n}\left[\frac{a}{t^{\sigma}}\left|\begin{array}{c}(a_p, A_p),(\rho,\sigma)\\(b_q, B_q)\end{array}\right.\right],$$

$$(10.40)$$

which immediately yields the marginal pdf $P_1(x,t)$,

$$P_1(x,t) = \frac{1}{4-\bar{\nu}}\frac{D_{\frac{1}{2}}^{-1}}{\Gamma\left(\frac{2-\bar{\nu}}{4-\bar{\nu}}\right)}\left(\frac{4D_{\frac{1}{2}}^{-1}}{(4-\bar{\nu})^2}\right)^{-\frac{1}{4-\bar{\nu}}}|x|^{\frac{1-\bar{\nu}}{2}}t^{-\frac{3-\bar{\nu}}{2(4-\bar{\nu})}}$$

$$\times H_{1,2}^{2,0}\left[\frac{4D_{\frac{1}{2}}^{-1}}{(4-\bar{\nu})^2}\frac{|x|^{\frac{4-\bar{\nu}}{2}}}{t^{1/2}}\left|\begin{array}{c}\left(\frac{5-\bar{\nu}}{2(4-\bar{\nu})},\frac{1}{2}\right)\\\left(\frac{1}{4-\bar{\nu}},1\right),\left(-\frac{1}{4-\bar{\nu}},1\right)\end{array}\right.\right].$$

$$(10.41)$$

Using the scaling property (2.41) and the Mellin transform (2.47) of the Fox function, we can show that the solution (10.41) is normalized,

$$\int_{-\infty}^{\infty}P_1(x,t)dx = 1.$$

$$(10.42)$$

The scaling property (2.41) yields for the Fox function in Eq. (10.41)

$$P_1(x,t) = B(D_{\frac{1}{2}},\bar{\nu},t)|x|^{\frac{1-\bar{\nu}}{2}}$$

$$\times H_{1,2}^{2,0}\left[\frac{4D_{\frac{1}{2}}^{-1}}{(4-\bar{\nu})^2}\frac{|x|^{\frac{4-\bar{\nu}}{2}}}{t^{1/2}}\left|\begin{array}{c}\left(\frac{5-\bar{\nu}}{2(4-\bar{\nu})},1/2\right)\\\left(\frac{1}{4-\bar{\nu}},1\right),\left(\frac{-1}{4-\bar{\nu}},1\right)\end{array}\right.\right]$$

$$= \frac{2B(D_{\frac{1}{2}},\bar{\nu},t)}{4-\bar{\nu}}|x|^{\frac{1-\bar{\nu}}{2}}$$

$$\times H_{1,2}^{2,0}\left[|x|\left(\frac{4D_{\frac{1}{2}}^{-1}t^{-\frac{1}{4-\bar{\nu}}}}{(4-\bar{\nu})^2}\right)^{\frac{4-\bar{\nu}}{2}}\left|\begin{array}{c}\left(\frac{5-\bar{\nu}}{2(4-\bar{\nu})},\frac{1}{4-\bar{\nu}}\right)\\\left(\frac{1}{4-\bar{\nu}},\frac{1}{4-\bar{\nu}}\right),\left(\frac{-1}{4-\bar{\nu}},\frac{1}{4-\bar{\nu}}\right)\end{array}\right.\right],$$

$$(10.43)$$

where $B(D_\nu,\bar{\nu},t)$ is the part of $P_1(x,t)$ that does not contain x and which reads

$$\frac{2B(D_{\frac{1}{2}},\bar{\nu},t)}{4-\bar{\nu}} = \frac{t^{-\frac{3-\bar{\nu}}{2(4-\bar{\nu})}}}{2\Gamma\left(\frac{2-\bar{\nu}}{4-\bar{\nu}}\right)}\left(\frac{4D_\nu^{-1}}{(4-\bar{\nu})^2}\right)^{\frac{3-\tilde{\nu}u}{4-\bar{\nu}}}.$$

$$(10.44)$$

Integration with respect to x corresponds to the Mellin transform, which yields

$$2t^{-\frac{3-\bar{\nu}}{2(4-\bar{\nu})}}\left(-\frac{4D_{\frac{1}{2}}^{-1}}{(4-\bar{\nu})^2}\right)^{\frac{3-\bar{n}u}{4-\bar{\nu}}}\Gamma\left(\frac{2-\bar{\nu}}{4-\bar{\nu}}\right). \tag{10.45}$$

Clearly, these two expressions compensate each other exactly, and eventually we arrive at the result of Eq. (10.42).

The solution (10.41) describes subdiffusion with the MSD

$$\langle x^2(t)\rangle = 2\int_0^\infty dx\, x^2 P_1(x,t) = \frac{\Gamma\left(\frac{8-\bar{\nu}}{4-\bar{\nu}}\right)}{\Gamma\left(\frac{2-\bar{\nu}}{4-\bar{\nu}}\right)}\left[\frac{4D_{\frac{1}{2}}^{-1}}{(4-\bar{\nu})^2}\right]^{-\frac{4}{4-\bar{\nu}}}t^{\frac{2}{4-\bar{\nu}}}, \tag{10.46}$$

which is obtained by repeating the same procedure used above to establish the normalization of Eq. (10.41).

The transport exponent in the MSD (10.46) varies over the range $\frac{1}{2}\leq\frac{2}{4-\bar{\nu}}<\frac{2}{3}$. The limiting case with $\bar{\nu}=0$ results in the continuous comb with the MSD $\langle x^2(t)\rangle\simeq t^{1/2}$. Note the absence of the normal diffusion limit for $\bar{\nu}=1$, since the description of the fractal fingers structure is strictly valid only for $b<l$, which implies $\bar{\nu}<1$.

10.5.2 *Superdiffusion due to the memory kernels*

We now analyze the effect of the memory kernels and consider a specific case with the kernels $\hat{\gamma}(s)=1$ and $\hat{\eta}(s)=s^{-1/2}$. This choice implies that the homogeneous part of the Green's function in Eq. (10.27) is given by

$$D_{\frac{1}{2}}^{-1}sr^{-\bar{\nu}/2}G(r,s)=\frac{\partial^2 G(r,s)}{\partial r^2}. \tag{10.47}$$

The only difference with the Bessel differential equation (10.30) is the power of the Laplace parameter. Instead of $\sqrt{s}$, we have s and the solution of Eq. (10.47) reads

$$G(|x|,s)=\sqrt{|x|}K_{\frac{2}{4-\bar{\nu}}}\left(\frac{4}{4-\bar{\nu}}D_{\frac{1}{2}}^{-1/2}s^{1/2}|x|^{\frac{4-\bar{\nu}}{4}}\right). \tag{10.48}$$

This provides the solution in the form of Eqs. (10.39) and (10.41) in the Laplace and time domains, respectively. However now the shift condition (10.28) at $r=0$ is

$$-2D_x\mathcal{C}_{\bar{\nu}}\frac{dG(r,s)}{dr}\bigg|_{r=0}=\sqrt{s}. \tag{10.49}$$

Therefore, following the same procedure as above, we find the marginal pdf,

$$P_1(x,t) = \frac{D_{\frac{1}{2}}^{-1}}{4-\bar{\nu}} \frac{1}{\Gamma\left(\frac{2-\bar{\nu}}{4-\bar{\nu}}\right)} \left(\frac{4D_{\frac{1}{2}}^{-1}}{(4-\bar{\nu})^2}\right)^{-\frac{1}{4-\bar{\nu}}} |x|^{\frac{1-\bar{\nu}}{2}} t^{-\frac{3-\bar{\nu}}{4-\bar{\nu}}}$$

$$\times H_{1,2}^{2,0}\left[\frac{4D_{\frac{1}{2}}^{-1}}{(4-\bar{\nu})^2}\frac{|x|^{\frac{4-\bar{\nu}}{2}}}{t}\left|\begin{array}{c}\left(\frac{1}{4-\bar{\nu}},\frac{1}{2}\right)\\\left(\frac{1}{4-\bar{\nu}},1\right),\left(-\frac{1}{4-\bar{\nu}},1\right)\end{array}\right.\right]. \quad (10.50)$$

As follows from the definition of the Fox function, we can use the reduction formula (2.42) for this result and find

$$P_1(x,t) = \frac{D_{\frac{1}{2}}^{-1}}{4-\bar{\nu}} \frac{1}{\Gamma\left(\frac{2-\bar{\nu}}{4-\bar{\nu}}\right)} \left(\frac{4D_{\frac{1}{2}}^{-1}}{(4-\bar{\nu})^2}\right)^{-\frac{1}{4-\bar{\nu}}} |x|^{\frac{1-\bar{\nu}}{2}} t^{-\frac{3-\bar{\nu}}{4-\bar{\nu}}}$$

$$\times H_{0,1}^{1,0}\left[\frac{4D_{\frac{1}{2}}^{-1}}{(4-\bar{\nu})^2}\frac{|x|^{\frac{4-\bar{\nu}}{2}}}{t}\left|\begin{array}{c}\underline{}\\\left(-\frac{1}{4-\bar{\nu}},1\right)\end{array}\right.\right]. \quad (10.51)$$

This Fox function corresponds to the stretched exponential function, which is a generalization of Example 2.7 in Chapter 2. For the special case of the parameters of the Fox H-function in Eq. (10.51), we obtain

$$H_{0,1}^{1,0}\left[z\left|\begin{array}{c}\underline{}\\(b,B)\end{array}\right.\right] = B^{-1}z^{b/B}\exp\left(-z^{1/B}\right). \quad (10.52)$$

This eventually yields the solution in elementary function form,

$$P_1(x,t) = \frac{D_{\frac{1}{2}}^{-1}}{4-\bar{\nu}} \frac{1}{\Gamma\left(\frac{2-\bar{\nu}}{4-\bar{\nu}}\right)} \left(\frac{4D_{\frac{1}{2}}^{-1}}{(4-\bar{\nu})^2}\right)^{-\frac{2}{4-\bar{\nu}}} |x|^{-\frac{\bar{\nu}}{2}} t^{-\frac{2-\bar{\nu}}{4-\bar{\nu}}}$$

$$\times \exp\left[-\frac{4D_{\frac{1}{2}}^{-1}}{(4-\bar{\nu})^2}\frac{|x|^{\frac{4-\bar{\nu}}{2}}}{t}\right]. \quad (10.53)$$

It is normalized to one and has indeed the form of a stretched exponential. This is a remarkable result. The MSD is finite and can be straightforwardly calculated for superdiffusion,

$$\langle x^2(t)\rangle = 2\int_0^\infty dx\, x^2 P_1(x,t) \simeq t^{\frac{4}{4-\bar{\nu}}}. \quad (10.54)$$

It describes superdiffusion with the transport exponent ranging over the interval $1 < \frac{4}{4-\bar{\nu}} < \frac{4}{3}$. The solution describes a Lévy-like process, where a CTRW with spatio-temporal coupling takes place. The diffusion in the x-direction is enhanced due to the generalized compensation memory kernel[6]

---

[6]The presence of this compensation memory kernel leads to normal diffusion in the x-direction in the limiting case of the continuous comb with $\bar{\nu} = 0$.

$\eta(t) = \frac{t^{-1/2}}{\Gamma(1/2)}$. The long jumps on the fractal comb are penalized by long waiting times. This mechanism leads to the stretched exponential behavior in Eq. (10.53), which yields the finite MSD. The case with $\bar{\nu} = 0$ recovers the result of the continuous comb with $\langle x^2(t) \rangle \simeq t$.

Concluding this section, we note that Lévy processes appear in the comb structure as the result of the fractional distribution of the fingers. A transition from subdiffusion to superdiffusion takes place due to the temporal memory kernels. We have provided exact analytical solutions, obtained in the form of the Fox H-functions. The fractional structure of the fingers is controlled by the Weierstrass function, which leads to a power-law kernel in Fourier space. The superdiffusive solution (10.50) is expressed in the form of a stretched exponential function (10.51). It is a special case, where the second moment exists for superdiffusion, since the Lévy flights are interrupted by the fingers-traps with a power-law waiting time pdf. In that case, the superdiffusive MSD is calculated exactly from the second moment, $\langle x^2(t) \rangle = t^{\frac{4}{4-\nu}}$. Here we have considered only one specific example of the spatio-temporal interplay for the specific realization of the system parameters. A variety of the spatio-temporal parameter with different realizations of anomalous kinetics is considered in Ref. [164].

10.6   Solution of the FFPE of a generalized comb

The general form of a space-time fractional diffusion equation contains both the Caputo time and the Riesz space fractional derivative, and it reads

$$\partial_t^\alpha P_1(x,t) = D_{\alpha,\beta} \frac{\partial^\beta P_1(x,t)}{\partial|x|^\beta}, \quad t > 0, \quad -\infty < x < +\infty. \tag{10.55}$$

Recall that ∂_t^α is the Caputo fractional time derivative of the order of $0 < \alpha < 1$, while $\frac{\partial^\beta}{\partial|x|^\beta}$ is the Riesz fractional space derivative of order $1 < \beta < 2$, and $D_{\alpha,\beta}$ is the generalized diffusion coefficient with physical dimension $[D_{\alpha,\beta}] = \mathrm{m}^\beta \mathrm{s}^{-\alpha}$. The initial and boundary conditions are

$$P_1(\pm\infty, t) = 0, \quad \frac{\partial}{\partial x} P_1(\pm\infty, t) = 0, \quad t > 0, \tag{10.56}$$

$$P_1(x, 0) = \delta(x), \quad -\infty < x < +\infty. \tag{10.57}$$

Applying the Fourier transformation to Eq. (10.55), we arrive at Example 2.16, where it was shown that the Mittag–Leffler function is a solution of the equation with the Caputo fractional derivative,

$$\partial_t^\alpha \tilde{P}_1(k,t) = D_{\alpha,\beta} |k|^\beta \tilde{P}_1(k,t). \tag{10.58}$$

The solution reads

$$\tilde{P}_1(k,t) = E_\alpha \left(-D_{\alpha,\beta} t^\alpha |k|^\beta\right). \tag{10.59}$$

Here $E_\lambda(z)$ is the one-parameter Mittag–Leffler function, which also can be represented as the Fox function

$$E_{\alpha,\beta}(-z) = H_{1,2}^{1,1} \left[z \, \middle| \, \begin{matrix} (0,1) \\ (0,1), (1-\beta,\alpha) \end{matrix} \right]. \tag{10.60}$$

Note that $\tilde{P}_1(k,0) = 1$.

The next step in the solution procedure is the inverse Fourier transformation in the form the Mellin-cosine transformation of the Fox H-function, which is given by the expression[7]

$$\int_0^\infty d\kappa \, \kappa^{\rho-1} \cos(\kappa x) H_{p,q}^{m,n} \left[a\kappa^\delta \, \middle| \, \begin{matrix} (a_p, A_p) \\ (b_q, B_q) \end{matrix} \right]$$

$$= \frac{\pi}{x^\rho} H_{q+1,p+2}^{n+1,m} \left[\frac{x^\delta}{a} \, \middle| \, \begin{matrix} (1-b_q, B_q), \left(\dfrac{1+\rho}{2}, \dfrac{\delta}{2}\right) \\ (\rho,\delta), (1-a_p, A_p), \left(\dfrac{1+\rho}{2}, \dfrac{\delta}{2}\right) \end{matrix} \right]. \tag{10.61}$$

Applying this Mellin-cosine transformation formula to Eq. (10.60) with $\rho = 1$, we have the following chain of transformations,

$$P_1(x,t) = \frac{1}{\pi} \int_0^\infty d k \, \cos(k\,x) H_{1,2}^{1,1} \left[D_{\alpha,\beta} t^\alpha |k|^\beta \, \middle| \, \begin{matrix} (0,1) \\ (0,1), (0,\alpha) \end{matrix} \right]$$

$$= \frac{1}{\beta\pi} \int_0^\infty d k \, \cos(k\,x) H_{1,2}^{1,1} \left[(D_{\alpha,\beta} t^\alpha)^{\frac{1}{\beta}} |k| \, \middle| \, \begin{matrix} (0,1/\beta) \\ (0,1/\beta), (0,\alpha/\beta) \end{matrix} \right]$$

$$= \frac{1}{\beta|x|} H_{3,3}^{2,1} \left[\frac{|x|}{(D_{\alpha,\beta} t^\alpha)^{\frac{1}{\beta}}} \, \middle| \, \begin{matrix} \left(1, \dfrac{1}{\beta}\right), \left(1, \dfrac{\alpha}{\beta}\right), \left(1, \dfrac{1}{2}\right) \\ (1,1), \left(1, \dfrac{1}{\beta}\right), \left(1, \dfrac{1}{2}\right) \end{matrix} \right]. \tag{10.62}$$

The calculation of the MSD poses a problem, since the second moment diverges due to properties of the gamma function. In this case, one calculates the fractional q moment [3]. From the solution (10.62), this procedure yields the Mellin transform of the Fox function (2.47),

$$\langle |x|^q(t) \rangle = \frac{2}{\beta} \int_0^\infty dx \, x^{q-1} H_{3,3}^{2,1} \left[\frac{|x|}{(D_{\alpha,\beta} t^\alpha)^{\frac{1}{\beta}}} \, \middle| \, \begin{matrix} \left(1, \dfrac{1}{\beta}\right), \left(1, \dfrac{\alpha}{\beta}\right), \left(1, \dfrac{1}{2}\right) \\ (1,1), \left(1, \dfrac{1}{\beta}\right), \left(1, \dfrac{1}{2}\right) \end{matrix} \right]$$

---

[7]The conditions of validity of the Mellin-cosine transform are defined in Eq. (2.54).

$$= \frac{2}{\beta} (D_{\alpha,\beta} t^{\alpha})^{\frac{q}{\beta}} \theta(-q)$$

$$= \frac{4}{\beta} \cdot \frac{\Gamma(q)\Gamma(1+q/\beta)\Gamma(-q/\beta)}{\Gamma(q/2)\Gamma(-q/2)\Gamma(1+\alpha q/\beta)} \cdot (D_{\alpha,\beta} t^{\alpha})^{\frac{q}{\beta}} . \tag{10.63}$$

10.7 Properties of memory kernels

In this section we focus on the memory kernels $\eta(t)$ and $\gamma(t)$. In the limit $\bar{\nu} = 0$, or $\nu = 1$, which corresponds to the continuously homogeneous distribution of the fingers, Eq. (10.19) becomes the following generalized comb-like model equation [164],

$$\int_0^t dt' \, \gamma(t-t') \frac{\partial}{\partial t'} P(x,y,t') = \mathcal{D}_x \delta(y) \int_0^t dt' \, \eta(t-t') \frac{\partial^2}{\partial x^2} P(x,y,t')$$

$$+ \mathcal{D}_y \frac{\partial^2}{\partial y^2} P(x,y,t) \tag{10.64}$$

with the same initial and boundary conditions as Eq. (1.1). This equation is the straightforward generalization of Eq. (8.24). As discussed at the beginning of this chapter, the memory kernels change the dimensionality of the diffusion coefficients. Therefore without loss of generality, we use dimensionless variables and set $\mathcal{D}_x = \mathcal{D}_y = 1$. Recall that $\gamma(t)$ and $\eta(t)$ are integrable non-negative memory kernels that approach zero in the long-time limit.[8] The memory kernel $\gamma(t)$ represents the memory effects of the system and means that particles moving along the fingers can also be trapped. Consequently, the diffusion along the y-direction can also be anomalous. Indeed, integration of Eq. (10.64) with respect to x yields the equation for the marginal pdf $P_2(y,t)$,

$$\int_0^t dt' \, \gamma(t-t') \frac{\partial}{\partial t'} P_2(y,t') = \frac{\partial^2}{\partial y^2} P_2(y,t), \tag{10.65}$$

---

[8]We shall discuss this property in more details below.

where $P_2(y,t) = \int_{-\infty}^{\infty} dx P(x,y,t)$. The memory kernel $\eta(t)$ was introduced to compensate the anomalous transport along the backbone.[9] We call the memory kernel $\eta(t)$ therefore a generalized compensation kernel.

As in previous sections, our main concern here is the marginal pdf $P_1(x,t)$. Integrating with respect to y, performing the Laplace and Fourier transformations[10] of Eq. (10.64), and introducing a new memory kernel

$$\xi(s) = \frac{1}{\eta(s)}\sqrt{\frac{\gamma(s)}{s}}, \tag{10.66}$$

we find the marginal pdf in the Fourier–Laplace domain,

$$\hat{P}_1(k,s) = \frac{\xi(s)}{s\xi(s) + \frac{1}{2}k^2}. \tag{10.67}$$

Comparing this result with the master equation (1.23) and the memory kernel $K(s) = s\psi(s)/[1-\psi(s)]$ in Eq. (1.24), we obtain the relation between the memory kernel $\xi(s)$ and the waiting time pdf,

$$\psi(s) = \frac{1}{1 + s\xi(s)}. \tag{10.68}$$

The normalization condition requires $\lim_{s\to 0} s\xi(s) = 0$, implying that $\xi(s)$ does not grow faster than s. This means that the diffusion equation

$$\int_0^t dt'\, \xi(t-t')\frac{\partial P_1(x,t')}{\partial t'} = \frac{1}{2}\frac{\partial^2 P_1(x,t)}{\partial x^2} \tag{10.69}$$

is the fractional time Fokker–Planck equation. From the solution (10.67) we also obtain the general form for the MSD,

$$\langle x^2(t)\rangle = \mathcal{L}^{-1}\left[-\frac{\partial^2}{\partial k^2}\hat{P}_1(k,s)\right]\Big|_{k=0} = \mathcal{L}^{-1}\left[\frac{1}{s^2\xi(s)}\right] = \mathcal{L}^{-1}\left[\frac{1}{s}\frac{\eta(s)}{\sqrt{s\gamma(s)}}\right]. \tag{10.70}$$

10.7.1 Subordination to the Wiener process

To further understand the properties of the kernels $\xi(s)$, $\eta(s)$, and $\gamma(s)$, we write the pdf (10.67) in integral form,

$$\hat{P}_1(k,s) = \xi(s)\int_0^{\infty} e^{-us\xi(s)}e^{-\frac{1}{2}uk^2}du. \tag{10.71}$$

---

[9]The fractional integral $I_t^{1-\gamma}$ plays this role in Eq. (8.24).
[10]We obtain the same result from the Fourier transformation of Eq. (10.25) with $\bar{\nu} = 0$ and $f(x,s)$ replaced by $\hat{P}_1(x,s)$.

After the inverse Laplace–Fourier transformation we arrive at the pdf that *subordinates* anomalous diffusion, governed by Eq. (10.67), on the time scale t (physical time) to the Wiener process (Brownian motion) on a time scale u (operational time) [51]. We obtain a scheme that represents the pdf $P_1(x,t)$ of a given random process $x(t)$ as follows,

$$P_1(x,t) = \int_0^\infty P_W(x,u)h_1(u,t)du, \qquad (10.72)$$

where

$$P_W(x,u) = \frac{1}{\sqrt{4\pi u}} \exp\left(-\frac{x^2}{4u}\right), \qquad (10.73)$$

is the Gaussian pdf, i.e., the pdf of the Wiener process, or Brownian motion, at the operational time $u < t$, and $h_1(u,t)$ is a pdf *subordinating* the random process $x(t)$ to the Wiener process.

We use Eq. (10.71) to define $h_1(u,t)$:

$$\hat{h}_1(u,s) \equiv \xi(s)e^{-us\xi(s)} = -\frac{\partial}{\partial u}\frac{1}{s}\hat{L}_1(s,u), \qquad (10.74)$$

where we introduce

$$\hat{L}_1(s,u) = e^{-us\xi(s)}. \qquad (10.75)$$

Note that the function $h_1(u,t)$ is normalized with respect to u. This immediately follows from Eq. (10.74),

$$\int_0^\infty du\, h_1(u,t) = \mathcal{L}_s^{-1}\left[\int_0^\infty du\, \xi(s)e^{-us\xi(s)}\right] = \mathcal{L}_s^{-1}\left[\frac{1}{s}\right] = 1. \qquad (10.76)$$

By using the subordination approach, we establish that the pdf $P_1(x,t)$ is non-negative if $h_1(u,t)$ is non-negative, as well. Therefore, it suffices to show that the Laplace image $\hat{h}_1(u,s)$ is a completely monotone function with respect to s. For this property to hold, both the functions $\xi(s) = \frac{1}{\eta(s)}\sqrt{\frac{\gamma(s)}{s}}$ and $e^{-us\xi(s)}$ should be completely monotone, as well. The function $e^{-us\xi(s)}$ is completely monotone, if $s\xi(s) = \frac{1}{\eta(s)}\sqrt{s\gamma(s)}$ is a Bernstein function, see Sec. 10.7.2.

We apply the same arguments to anomalous diffusion along the fingers, which is described by the marginal pdf $P_2(y,t)$ in Eq. (10.65) and find

$$P_2(y,t) = \int_0^\infty P_W(y,u)h_2(u,t)du, \qquad (10.77)$$

which defines $h_2(u,t)$ via its Laplace transform $\hat{h}_2(u,s)$,

$$\hat{h}_2(u,s) = \gamma(s)e^{-us\gamma(s)}. \qquad (10.78)$$

The function $\hat{h}_2(u, s)$ is a completely monotone function, if the function $\gamma(s)$ is completely monotone and $s\gamma(s)$ is a Bernstein function. These conditions provide restrictions on the possible choices for the kernels $\gamma(t)$ and $\eta(t)$ in the generalized comb model (10.64).

To conclude, we note that here the comb geometry does not play the dominant role for the CTRW, since the latter is ultimately determined by the memory kernels $\eta(t)$ and $\gamma(t)$. However the comb remains a useful and important geometry to study in this setting due to its occurrence in natural systems, like the dendritic spines studied in Chapter 8, or in experimental realizations, like the microfluidic device studied in Chapter 9, where the anomalous transport takes place in the comb geometry.

10.7.2 Auxiliary: Completely monotone and Bernstein functions

For the convenience of the reader, we provide a short list of definitions and properties of completely monotone functions and Bernstein functions, mainly based on Ref. [168].

(1) The function $g(s)$ is completely monotone if $(-1)^n g^{(n)}(s) \geq 0$ for all $n \geq 0$ and $s > 0$. A product of completely monotone functions is a completely monotone function too. An example of a completely monotone function is s^α, where $\alpha < 0$.

(2) The Bernstein characterization theorem states that if the Laplace transform $g(s) = \mathcal{L}[G(t)]$ of a given function $G(t)$ is a completely monotone function, then the function $G(t)$ is non-negative.

(3) A given function $f(s)$ is a Bernstein function if $(-1)^{n-1} f^{(n)}(s) \geq 0$ for all $n \in N$ and $s > 0$. An example of a Bernstein function is s^α, where $0 < \alpha < 1$.

(4) All Bernstein functions have a unique Lévy–Khintchine representation,

$$f(s) = a + bs + \int_{(0,\infty)} \left(1 - e^{-sx}\right) \mu(dx), \qquad (10.79)$$

where $a, b \geq 0$ and μ is a measure on $(0, \infty)$ satisfying $\int_{(0,\infty)} (x \wedge 1) \mu(dx) < \infty$ [169].

(5) A Bernstein function $f(s)$ is a complete Bernstein function if its Lévy measure μ in (10.79) has a completely monotone density m with respect to the Lebesgue measure [169].

(6) The function $e^{-us\xi(s)}$ is completely monotone, if $s\xi(s)$ is a complete Bernstein function.

(7) If $f(s)$ is a complete Bernstein function, then $g(s) = 1/f(s)$ is a completely monotone function [170].

(8) A function $f(s)$ is a complete Bernstein function if and only if the function $s/f(s)$ is complete Bernstein function.

Chapter 11

Quantum comb model

11.1 Introduction

In quantum physics, the fractional concept can be introduced by means of the Feynman propagator and the Feynman path integral formulation of non-relativistic quantum mechanics [171]. The equivalence between the Wiener and Feynman path integrals, established by Kac [172], indicates a relation between the classical diffusion equation and the Schrödinger equation. Space fractional derivatives in the Schrödinger equation appear naturally, as shown in the seminal papers [173–177], when quantum mechanics is based on a path integrals approach. Replacing the Wiener functional measure, which is related to the process of Brownian motion, by a functional measure related to infinitely divisible statistical distributions, such as Lévy stable processes, corresponds to replacing the classical diffusion equation by the space fractional diffusion equation. Similarly, replacing the Feynman functional measure by a quantum Feynman–Lévy measure corresponds to replacing the standard Schrödinger equation by a *space fractional* Schrödinger equation [173–177].

In the context of diffusion, there exist various extensions of the classical diffusion equation to fractional diffusion equations which describe anomalous diffusion, see, e.g., Ref. [178]. The space fractional diffusion equation still describes a Markov process. The time fractional diffusion equation describes a non-Markovian process and can account for subdiffusion and superdiffusion. Additionally, there exists the mixed case with both space and time fractional derivatives. The *time fractional* Schrödinger equation (TFSE) was first considered by Naber [178], where a fractional time derivative was introduced in analogy with the fractional diffusion or Fokker–Planck equation, by means of the Wick rotation of time, $t \to t/i\hbar$. The

dynamics described by the TFSE does not correspond to a unitary transformation. The Green's function is found in the form of the Mittag–Leffler
function and does not satisfy Stone's theorem on one-parameter unitary
groups [179].

The time fractional quantum dynamics with the Hamiltonian $\mathcal{H}(x)$ is
described by the TFSE[1]

$$(i\hbar_{\text{eff}})^\alpha \frac{\partial^\alpha \Psi(x,t)}{\partial t^\alpha} = \mathcal{H}\Psi(x,t), \tag{11.1}$$

where $\alpha \leq 1$. Care must be taken to preserve the units of the wave function
when a fractional time derivative is introduced, and therefore all variables
and parameters in Eq. (11.1) are considered dimensionless[2] and $\hbar_{\text{eff}}$ is the
dimensionless Planck constant. For $\alpha = 1$, Eq. (11.1) is the "conventional"
(standard) Schrödinger equation. For $\alpha < 1$, the fractional derivative is a
formal notation of the Caputo fractional derivative.

11.2   Quantum comb model and TFSE

The quantum behavior of a particle on the comb can be defined as the
quantum motion in the $d + 1$ configuration space $(\mathbf{x}, y)$, such that the dynamics in the d-dimensional configuration space $\mathbf{x}$ is possible only at $y = 0$,
and motions in the $\mathbf{x}$ and y directions commute. Therefore the quantum
dynamics is described by the following Schrödinger equation,

$$i\hbar_{\text{eff}} \frac{\partial \Psi}{\partial t} = \delta(y)\mathbf{H}(\mathbf{x})\Psi - \frac{\hbar_{\text{eff}}^2}{2}\frac{\partial^2 \Psi}{\partial y^2}, \tag{11.2}$$

where the Hamiltonian is $\mathbf{H} = -\frac{\hbar_{\text{eff}}^2}{2}\nabla + V(\mathbf{x})$. It governs the dynamics of
a particle in a potential $V(\mathbf{x})$ in the $\mathbf{x}$-space, while the y coordinate corresponds to one-dimensional free motion. All the parameters and variables
are dimensionless.[3]

In analogy with diffusion in the comb model (1.1), we focus on the
dynamics in the $\mathbf{x} \equiv x$ space. (For simplicity of notation, we replace $\mathbf{x}$ by

---

[1]One should not confuse the wave function $\Psi(x,t)$, which is a function of the coordinates
x, with the cumulative probability $\Psi(t)$, considered in the previous chapters, which is a
function of time only.

[2]Following Ref. [178], we introduce the Planck length $L_P = \sqrt{\hbar G/c^3}$, time $T_P = \sqrt{\hbar G/c^5}$, mass $M_P = \sqrt{\hbar c/G}$, and energy $E_P = M_P c^2$, where $\hbar$, G, and c are the
Planck constant, the gravitational constant, and the speed of light, respectively.

[3]In analogy with the TFSE (11.1), the quantum mechanics of a particle with mass
m is described by the dimensionless units $x/L_P \to x$, $y/L_P \to y$, $t/T_P \to t$, while the
dimensionless Planck constant is defined as the inverse dimensionless mass $\hbar_{\text{eff}} = M_P/m$.
Note that the dimensionless potential is now $V(x) \to V(x)/M_P c^2$, see also Footnote 2.

x.) However, a simple integration of the wave function with respect to y is not valid. It has no physical meaning, since it destroys the marginal pdf, $P_1(x, t) = \int |\Psi(x, y, t)|^2 dy \neq |\int \Psi(x, y, t) dy|^2$. To better understand the role of the fractional time derivative, we perform the Fourier transformation with respect to y, denoting the conjugate Fourier variable by l,

$$\mathcal{F}_y \Psi(x, y, t) = \tilde{\Psi}(x, l, t) \equiv \Psi_l(x, t) = \Psi_l, \tag{11.3}$$

and Eq. (11.2) reads,

$$i\hbar_{\text{eff}} \frac{\partial \Psi_l}{\partial t} = \mathbf{H}(x)\Psi(x, 0, t) + \frac{\hbar_{\text{eff}} l^2}{2} \Psi_l. \tag{11.4}$$

To obtain this equation in closed form, we express the wave function at $y = 0$, $\Psi(x, 0, t)$, in terms of the Fourier transform Ψ_l. Following the standard procedure, we perform the Laplace transformation of Eq. (11.2) with respect to time, which yields the solution in the Laplace domain,

$$\hat{\Psi}(x, y, s) = \hat{\Psi}(x, 0, s) \exp\left[-\sqrt{2s/i\hbar_{\text{eff}}}|y|\right]. \tag{11.5}$$

Applying the Fourier transformation with respect to y, we obtain from Eq. (11.5)

$$\hat{\Psi}_l(x, s) = \hat{\Psi}(x, y = 0, s)\mathcal{F}_y e^{-\sqrt{2s/i\hbar_{\text{eff}}}|y|} = \frac{2(1 - i)\sqrt{s/\hbar_{\text{eff}}}}{l^2 - 2is/\hbar_{\text{eff}}} \hat{\Psi}(x, y = 0, s), \tag{11.6}$$

where $1 - i = \sqrt{2/i}$. Simple algebra leads to

$$\hat{\Psi}(x, 0, s) = \left[\hat{\Psi}_l(x, s) \frac{l^2 - 2is/\hbar_{\text{eff}}}{2(1 - i)\sqrt{s/\hbar_{\text{eff}}}}\right], \tag{11.7}$$

and the inverse Laplace transformation of $\hat{\Psi}(x, 0, s)$ yields the wave function at $y = 0$. Consider the simple case with $l = 0$. We have from Eq. (11.7)

$$\hat{\Psi}_0(x, 0, s) = \sqrt{\frac{s}{2\hbar_{\text{eff}} i}} \hat{\Psi}_0(x, s). \tag{11.8}$$

Then carrying out the Laplace transformation in Eq. (11.4), we arrive at the definition of the Caputo fractional derivative $\partial_t^{\frac{1}{2}}$ with the property

$$\mathcal{L}[\partial_t^{1/2}] = s^{1/2}\hat{\Psi}_0(x, s) - s^{-1/2}\Psi_0(x, t = 0). \tag{11.9}$$

Carrying out the inverse Laplace transformation and redefining $\mathbf{H}/\sqrt{2} \to \mathcal{H}$, we obtain a TFSE that coincides exactly with Eq. (11.1) with $\alpha = 1/2$, and which can be obtained via the Wick rotation from the FFPE (1.17),

$$\sqrt{i\hbar_{\text{eff}}}\partial_t^{\frac{1}{2}}\Psi_0(x, t) = \mathcal{H}\Psi_0(x, t). \tag{11.10}$$

If we neglect all nonzero modes, then the information contained in $l \neq 0$ is lost. Therefore, Eq. (11.10) is simply an incomplete description of the quantum mechanics of a particle in the x-space, which interacts with the y-space environment, where most information about the environment and the interaction is neglected. Repeating these simple operations with the Laplace transformation for the $l \neq 0$ modes, we obtain the complete and exact TFSE of the quantum comb,

$$(i\hbar_{\text{eff}})^{\frac{1}{2}}\partial_t^{\frac{1}{2}}\Psi_l = -\frac{i\hbar_{\text{eff}}l^2}{2\sqrt{2}}\mathcal{D}_t^{-1}\mathbf{H}(x)\Psi_l + \frac{\mathbf{H}(x)}{\sqrt{2}}\Psi_l + (i\hbar_{\text{eff}})^{\frac{1}{2}}\mathcal{D}_t^{-\frac{1}{2}}\frac{\hbar_{\text{eff}}^2 l^2}{2}\Psi_l.$$

$$(11.11)$$

This comb TFSE describes the quantum dynamics in the $\mathbf{x}$ configuration space. The subscript l corresponds to an effective interaction of a quantum system with an additional degree of freedom, while the fractional time derivatives, with $\alpha = 1/2$, reflect this interaction in the form of non-Markovian memory effects. The TFSE (11.10) is an approximation of the non-Hermitian (non-unitary) dynamics, which is valid when the amplitudes of the l-th modes are negligibly small.

11.3   Green's function

The initial value problem with the initial condition $\Psi(t = 0) = \Psi_0(x, y)$ is described by Green's function. In complete analogy with the diffusive comb model and fractional diffusion, the boundary conditions are the same in the y-configuration subspace. We can also define the eigenvalue problem for the Hamiltonian,

$$\mathbf{H}(x)\psi_\lambda(x) = \lambda\psi_\lambda(x), \qquad (11.12)$$

and write the wave function in Eq. (11.2) as the series

$$\Psi(x, y, t) = \sum_\lambda \phi_\lambda(y, t)\psi_\lambda(x), \qquad (11.13)$$

where $\sum_\lambda$ is understood as the integration with respect to λ for a continuous spectrum. For fixed λ we arrive at the dynamics of a particle in the δ potential,

$$i\hbar_{\text{eff}}\frac{\partial\phi_\lambda}{\partial t} = -\frac{\hbar_{\text{eff}}^2}{2}\frac{\partial^2\phi_\lambda}{\partial y^2} + \lambda\delta(y)\phi_\lambda. \qquad (11.14)$$

Taking into account that the Green's function of Eq. (11.2) has the spectral decomposition,

$$G(x, y, t; x'y') = \sum_\lambda G_\lambda(y, t; y')\psi_\lambda^*(x)\psi_\lambda(x'), \qquad (11.15)$$

we obtain the Schrödinger equation for the Green's function $G_\lambda = G_\lambda(y, t; y' = 0)$,

$$i\hbar_{\text{eff}}\frac{\partial G_\lambda}{\partial t} = -\frac{\hbar_{\text{eff}}^2}{2}\frac{\partial^2 G_\lambda}{\partial y^2} + \lambda\delta(y)G_\lambda + i\hbar_{\text{eff}}\delta(y)\delta(t). \tag{11.16}$$

Here the initial condition is already taken into account. The Green's function for this Schrödinger equation has been obtained in Refs. [180, 181] for free boundary conditions at infinity.[4] For the chosen zero boundary conditions, it is instructive to employ Eq. (11.5) in Laplace space. Replacing the eigenvalues λ by the Hamiltonian $\mathbf{H}(\mathbf{x}) \equiv \mathbf{H}(x)$, we obtain the Green's function in the form of the inverse Laplace transform,

$$G\Big[\mathbf{H}(x), y, t\Big] = \mathcal{L}^{-1}\left[\frac{\hbar_{\text{eff}}}{i}\frac{e^{-|y|\sqrt{2s/i\hbar_{\text{eff}}}}}{\mathbf{H}(x) + \sqrt{2s\hbar_{\text{eff}}^3/i}}\right]. \tag{11.17}$$

We perform the inverse Laplace transformation, using the following representation for the denominator,

$$\frac{1}{\mathbf{H}(x) + \sqrt{s\hbar_{\text{eff}}^3/i}} = \int_0^\infty \exp\left\{-u[\mathbf{H}(x) + \sqrt{2s\hbar_{\text{eff}}^3/i}]\right\}du. \tag{11.18}$$

This representation is valid for any spectrum λ of the Hamiltonian due to the second term in the exponential and yields the integration

$$G\Big[\mathbf{H}(x), y, t\Big] = \frac{\hbar_{\text{eff}}}{i}\int_0^\infty \exp\left[-u\mathbf{H}(x)\right]$$

$$\times \left\{\frac{1}{2\pi i}\int_{-i\infty+c}^{i\infty+c} \exp(st)\exp\left[-\sqrt{\frac{2s/i}{\hbar_{\text{eff}}}}(|y| + \hbar_{\text{eff}}^2 u)\right]ds\right\}du. \tag{11.19}$$

After a simple manipulation, the inverse Laplace transformation reduces to a Gaussian integration,

$$\frac{1}{2\pi i}\int_{-i\infty+c}^{i\infty+c} \exp(st)\exp\left[-\sqrt{\frac{2s/i}{\hbar_{\text{eff}}}}(|y| + \hbar_{\text{eff}}^2 u)\right]ds$$

$$= \frac{(|y| + \hbar_{\text{eff}}^2 u)}{\sqrt{2i\hbar_{\text{eff}}\pi t^3}}\exp\left[-\frac{(|y| + \hbar_{\text{eff}}^2 u)^2}{2\hbar_{\text{eff}}it}\right]. \tag{11.20}$$

Therefore, the Green's function reads

$$G\Big[\mathbf{H}(x), y, t\Big] = \frac{\sqrt{i\hbar_{\text{eff}}}}{\sqrt{2\pi t^3}}\int_0^\infty (|y| + \hbar_{\text{eff}}^2 u)\exp\left[-u\mathbf{H}(x) - \frac{i(|y| + \hbar_{\text{eff}}^2 u)^2}{2\hbar_{\text{eff}}t}\right]du. \tag{11.21}$$

---

[4]We note that a more general situation is studied Ref. [180].

Using the Fourier transformation of the exponential,

$$\exp\left[\frac{i(|y| + \hbar_{\text{eff}}^2 u)^2}{2\hbar_{\text{eff}}t}\right]$$

$$= \sqrt{\frac{\hbar_{\text{eff}}t}{2\pi i}} \int_{-\infty}^{\infty} \exp\left(i\hbar_{\text{eff}}t\xi^2/2\right) \exp\left[-i\xi(|y| + \hbar_{\text{eff}}^2 u)\right] d\xi, \qquad (11.22)$$

we represent the Green's function in the following form,

$$G(x, y, t; x') = \frac{h}{2\pi t} \int_{-\infty}^{\infty} \left\{ \exp\left(i\hbar_{\text{eff}}t\xi^2/2\right) \left(i\frac{\partial}{\partial\xi}\right) \right.$$

$$\left. \times \int_0^{\infty} \exp\left[-u\mathbf{H}(x) - i\xi(|y| + \hbar_{\text{eff}}^2 u)\right] du \right\} d\xi. \qquad (11.23)$$

This expression is convenient for further analysis in the framework of path integrals.

11.4   Path integral representation

As an example, it is instructive to consider a free particle, because it has a straightforward relation to the original diffusive comb model (1.1). We find the Green's function along the structure x-axis in the coordinate space, $G(x, y, t; x') = \langle x'|G[\mathbf{H}(x), y, t]|x\rangle$, for a free particle of unit mass with the Hamiltonian $\mathbf{H} = p^2/2$. Expressing the exponential $\langle x'|e^{-u\mathbf{H}}|x\rangle$ in path integral form,

$$\langle x'|e^{-u\mathbf{H}}|x\rangle = \frac{1}{\sqrt{2\pi\hbar_{\text{eff}}^2 u}} \exp\left[-\frac{(x - x')^2}{2u\hbar_{\text{eff}}^2}\right], \qquad (11.24)$$

we obtain from Eq. (11.23):

$$G(x, y, t; x') = \frac{1}{t\sqrt{(2\pi)^3}} \int_{-\infty}^{\infty} \left\{ \exp\left(i\hbar_{\text{eff}}t\xi^2/2\right) \left(i\frac{\partial}{\partial\xi}\right) \right.$$

$$\left. \times \int_0^{\infty} \exp\left[-\frac{(x - x')^2}{2\hbar_{\text{eff}}^2 u} - i\xi(|y| + \hbar_{\text{eff}}^2 u)\right] \frac{du}{\sqrt{u}} \right\} d\xi. \qquad (11.25)$$

Integration with respect to the complex "time" u yields the following expression,

$$I(\sigma, \varsigma) = \int_0^{\infty} \exp\left[-\frac{\sigma}{u} - \varsigma u\right] \frac{du}{\sqrt{u}}, \qquad (11.26)$$

where $\sigma = (x - x')^2/(2\hbar_{\text{eff}}^2)$ and $\varsigma = i\xi\hbar_{\text{eff}}^2$. Differentiation of Eq. (11.26) with respect to σ and ς yields the equation

$$\frac{\partial^2 I(\sigma, \varsigma)}{\partial\varsigma\partial\sigma} = I(\sigma, \varsigma). \qquad (11.27)$$

By straightforward substitution we verify that the solution of this equation is

$$I(\sigma, \varsigma) = \sqrt{\frac{\pi}{\varsigma}} \exp\left[2\sqrt{-\sigma\varsigma}\right], \tag{11.28}$$

which also satisfies the "boundary" condition $I(0, \varsigma) = \sqrt{\pi/\varsigma}$. Having performed the integration with respect to u, we need to integrate with respect to ξ, which is carried out in the stationary phase approximation for the long time asymptotics $\hbar_{\mathrm{eff}} t \gg 1$,

$$\int_{-\infty}^{\infty} F(\xi) \exp[i\hbar_{\mathrm{eff}} t \xi^2/2 - i\xi|y|]d\xi \approx \sqrt{\frac{\pi}{i\hbar_{\mathrm{eff}} t}} F(\xi_0) \exp\left[-i\frac{y^2}{4\hbar_{\mathrm{eff}} t}\right], \tag{11.29}$$

where the stationary point is $\xi_0 = \frac{|y|}{\hbar_{\mathrm{eff}} t}$ and

$$\begin{aligned}
F(\xi) &= \sqrt{\frac{\pi}{i\hbar_{\mathrm{eff}}^2}} \left[\frac{-i}{2}\xi^{-\frac{3}{2}} - i(1+i)|x - x'|\xi^{-2} + |y|\xi^{-\frac{1}{2}}\right] \\
&\quad \times \exp\left(-\sqrt{2i\xi}|x - x'|\right) \\
&= \sqrt{\frac{\pi}{i\hbar_{\mathrm{eff}}^2}} F_0(\xi) \exp\left(-\sqrt{2i\xi}|x - x'|\right), \tag{11.30}
\end{aligned}$$

where $F_0(\xi)$ equals the square brackets expression. Taking the integration (11.29) into account, we eventually arrive at the Green's function

$$\begin{aligned}
G(x, y, t; x') &\approx \frac{-1}{4\sqrt{\pi\hbar_{\mathrm{eff}}^2 t^3}} F_0\left(\frac{|y|}{\hbar_{\mathrm{eff}} t}\right) \\
&\quad \times \exp\left[-i\frac{y^2}{\hbar_{\mathrm{eff}} t} - (1+i)|x - x'|\sqrt{|y|/\hbar_{\mathrm{eff}} t}\right]. \tag{11.31}
\end{aligned}$$

This solution also satisfies the boundary conditions at $x, y = \pm\infty$, where the Green's function vanishes.

11.5 Properties of the Green's function

Let us return to Eq. (11.20) and represent the result of the inverse Laplace transformation as follows,

$$\frac{(|y| + \hbar_{\mathrm{eff}}^2 u)}{\sqrt{2i\hbar_{\mathrm{eff}} \pi t^3}} \exp\left[-\frac{(|y| + \hbar_{\mathrm{eff}}^2 u)^2}{2\hbar_{\mathrm{eff}} it}\right] = \lim_{\xi \to 1} \frac{d}{d\xi} \frac{-\sqrt{i}}{|y|\sqrt{\pi t}} \exp\left[-\frac{(\xi|y| + \hbar_{\mathrm{eff}}^2 u)^2}{2\hbar_{\mathrm{eff}} it}\right]. \tag{11.32}$$

Therefore, the final step of the integrations in Eq. (11.19) is the Gaussian integral with respect to u,

$$\int_0^\infty \exp\left[\frac{i\hbar_{\text{eff}}^3}{2t}u^2 - \bar{\mathbf{H}}(\xi)u\right] du$$

$$= \frac{\sqrt{i\pi t}}{2\hbar_{\text{eff}}^3} \exp\left[\frac{2it}{\hbar_{\text{eff}}^3}\bar{\mathbf{H}}^2(x,|y|)\right] \text{erfc}\left[\sqrt{\frac{2it}{\hbar_{\text{eff}}^3}}\bar{\mathbf{H}}(\xi)\right], \qquad (11.33)$$

where $\bar{\mathbf{H}}(\xi) \equiv \bar{\mathbf{H}}(x,\xi|y|) = \mathbf{H}(x) - i\xi\frac{\hbar_{\text{eff}}|y|}{2t}$ is the effective Hamiltonian, which accounts for a non-Hermitian interaction with the fingers, and erfc is the complementary error function [46]. Collecting the results of Eqs. (11.32) and (11.33), we obtain from Eq. (11.19)

$$G\left[\mathbf{H}(x,y,t)\right] = \frac{-1}{2|y|}\lim_{\xi\to 1}\frac{d}{d\xi}\exp\left[\frac{2it}{\hbar_{\text{eff}}^3}\bar{\mathbf{H}}^2(\xi)\right]\text{erfc}\left[\sqrt{\frac{2it}{\hbar_{\text{eff}}^3}}\bar{\mathbf{H}}(\xi)\right]. \quad (11.34)$$

We now discuss some properties of this Green's function.

It satisfies zero boundary conditions at $|y| \to \infty$. The asymptotic behavior of the complementary error function is

$$\text{erfc}(z) \sim e^{-z^2}/z. \qquad (11.35)$$

Since $|y|$ is the dominant variable, we define $z = \xi|y|$ and obtain an approximation for the Green's function,

$$G\left[\mathbf{H}(x,y,t)\right]$$

$$\propto \frac{1}{|y|}\lim_{\xi\to 1}\frac{d}{d\xi}\left\{\exp\left(\frac{-i}{2t\hbar_{\text{eff}}}\xi^2 y^2\right)\text{erfc}\left[\left(\frac{-i}{2t\hbar_{\text{eff}}}\right)^{\frac{1}{2}}\xi|y|\right] + O\left(\frac{\mathbf{H}^2}{y^2}\right)\right\}$$

$$\sim \frac{-i}{y^2}\xrightarrow{y\to\pm\infty} 0. \qquad (11.36)$$

Let us express the result in Eq. (11.33) in terms of the Mittag–Leffler function. We now define z as

$$-iz = \sqrt{\frac{2it}{\hbar_{\text{eff}}^3}}\bar{\mathbf{H}}(\xi). \qquad (11.37)$$

Then the result in Eq. (11.34) can be rewritten in the form of the complex error function [46],[5]

$$w(z) = e^{-z^2}\text{erfc}(-iz). \qquad (11.38)$$

We employ the integral representation of $w(z)$ [46], and to this end we consider the Fourier transform[6] of the convolution integral

$$\tilde{w}(k) = \frac{1}{i\pi}\mathcal{F}\left[\int_{-\infty}^\infty \frac{e^{-u^2}du}{u-z}\right](k) = 2\theta(k)e^{-k^2/4}, \qquad (11.39)$$

---

[5]In Ref. [46] $w(z)$ is unnamed.
[6]Here we used the Fourier transformation as defined in the mathematical literature, $\tilde{f}(k) = \int_{-\infty}^\infty f(x)e^{-ikx}dx$, to have correspondence with Ref. [46].

where $\theta(k)$ is the Heaviside step function. Note that the Heaviside $\theta(k)$ function in the inverse Fourier transformation truncates the integration,

$$w(z) = \frac{1}{\pi} \int_{-\infty}^{\infty} \theta(k) e^{-\frac{k^2}{4} + ikz} dk = \frac{2}{\pi} \int_{0}^{\infty} e^{-u^2 + 2iuz} du = e^{-z^2} \operatorname{erfc}(-iz), \tag{11.40}$$

which yields the complex error function in Eq. (11.38).

Using this integral representation of $w(z)$ and performing the variable change $u^2 = -s$, we find

$$w(z) = \frac{1}{i\pi} \int_{-\infty}^{\infty} \frac{e^{-u^2} du}{u - z} = \frac{1}{2\pi i} \int_{-i\infty}^{i\infty} \frac{e^s s^{-\frac{1}{2}} ds}{s^{\frac{1}{2}} + iz} \tag{11.41}$$

$$= \sum_{n=0}^{\infty} \frac{(iz)^n}{\Gamma\left(\frac{1}{2}n + 1\right)} = E_{\frac{1}{2}}(iz), \tag{11.42}$$

where the last integration in Eq. (11.41) is considered as the inverse Laplace transformation with a corresponding deformation of the integration contour. Therefore, we have established the relation between the complex error function and the Mittag–Leffler function of the complex argument.

As follows from the integral representation of the Mittag–Leffler function in Eq. (11.41), the latter can be represented as a sum of the residues at $s \neq 0$ and the integral along the Hankel contour (in the negative direction), which accounts for the branch point at $s = 0$. These are oscillatory and decay terms, respectively,

$$E_{\frac{1}{2}}(z) = \frac{1}{2\pi i} \int_{-i\infty}^{i\infty} \frac{e^s s^{-\frac{1}{2}} ds}{s^{\frac{1}{2}} + iz} = \text{Residue} - \frac{1}{2\pi i} \int_{\text{Ha}} \cdots . \tag{11.43}$$

This kind of evaluation of the inverse Laplace transformation (11.41) has been employed by Naber [178], and this form is also used for the estimation of the large argument asymptotics of the Mittag–Leffler function [43].

First, let us evaluate the Residue,

$$\text{Residue} = \lim_{s \to -z^2} \frac{(s + z^2) e^s s^{-\frac{1}{2}}}{s^{\frac{1}{2}} + iz} = 2e^{-z^2}, \tag{11.44}$$

while the integral along the Hankel contour from $-\infty$ around $s = 0$ and back to $-\infty$ with a branch cut along the negative real axis reads

$$\frac{1}{2\pi i} \int_{\text{Ha}} \cdots = \frac{-iz}{\pi} \int_{0}^{\infty} \frac{e^s s^{-\frac{1}{2}} ds}{s - z^2}. \tag{11.45}$$

11.6 Nonlinear backbone dynamics

Using the properties of the Green's function established above, we can rewrite it in the form of the Mittag–Leffler function as follows,

$$G\left[\mathbf{H}(x), y, t\right] = \frac{i\hbar_{\text{eff}}}{2t} \frac{d}{d\mathbf{H}} E_{\frac{1}{2}}\left(-\bar{\mathbf{H}}\sqrt{\frac{2it}{\hbar_{\text{eff}}^3}}\right), \qquad (11.46)$$

where the limit $\xi = 1$ has already been taken. Now we can consider the quantum dynamics on the backbone at $y = 0$. In the light of the results obtained in Eqs. (11.43), (11.44) and (11.45), we retain the oscillating part of the Green's function and disregard the decay term. The "truncated" Green's function G_T is

$$G_T\left(\mathbf{H}, t\right) = \frac{i\hbar_{\text{eff}}}{t} \frac{d}{d\mathbf{H}} \exp\left(-\frac{2it\mathbf{H}^2}{\hbar_{\text{eff}}^3}\right) = \frac{2\mathbf{H}}{\hbar_{\text{eff}}^2} \exp\left(-\frac{2it\mathbf{H}^2}{\hbar_{\text{eff}}^3}\right). \qquad (11.47)$$

As an example, we study the quantum dynamics near a hyperbolic point. The system is described by the *linear* Hamiltonian[7]

$$\mathbf{H} = -i\hbar_{\text{eff}}x\frac{d}{dx} - \frac{i\hbar_{\text{eff}}}{2}. \qquad (11.48)$$

However, the backbone dynamics governed by the evolution operator (11.47) is completely nonlinear, and this leads to essential singularities of the evolution of an initial wave function $\Psi_0(x)$, namely $\Psi(x, t) = G_T\Psi_0(x)$ can bring a quantum particle to infinity during a finite time [184, 185]. Therefore, the formal treatment of this nonlinear dynamics needs some care [185].

Our main aim here is to evaluate the MSD,

$$\langle x^2(t)\rangle = \langle \Psi(x,t)\left|x^2\right|\Psi(x,t)\rangle = \int_{-\infty}^{\infty} x^2|\Psi(x,t)|^2 dx. \qquad (11.49)$$

11.6.1 *Eigenvalue problem*

First we consider the dimensionless equation for the eigenfunctions $\chi_\epsilon(x)$,

$$\frac{\hbar_{\text{eff}}}{i}\left(x\frac{d}{dx} + \frac{1}{2}\right)\chi_\epsilon(x) = \epsilon\chi_\epsilon(x) \qquad (11.50)$$

with the solution

$$\chi_\epsilon(x) = \frac{1}{\sqrt{N|x|}} \exp\left[\frac{i\epsilon}{\hbar_{\text{eff}}}\ln|x|\right], \qquad (11.51)$$

[7]It is one of the simple paradigms related to the Riemann hypothesis [182, 183].

where ϵ is the spectrum, and $N = 4\pi\hbar_{\text{eff}}$ is the normalization constant. The solution (11.51) satisfies the boundary conditions $\chi_\epsilon(x = \pm\infty) = 0$. For the continuous spectrum, the normalization condition is

$$\int_{-\infty}^{\infty} \chi_{\epsilon'}^*(x)\chi_\epsilon(x)dx = \delta(\epsilon - \epsilon'). \qquad (11.52)$$

A mathematically rigorous calculation of the normalization constant for the wave function $\chi_\epsilon(x)$ can be achieved by following the discussion in the monograph by V.A. Fock [186]. Since the operator $ix\frac{d}{dx}$ has a continuous spectrum ϵ, the eigenfunctions $\chi(\epsilon, x) \equiv \chi_\epsilon(x)$ are not square integrable. Therefore, the normalization condition exists not for the eigenfunction but for the "eigendifferential" $\Delta\chi(\epsilon, x)$ [186], which reads

$$\Delta\chi(\epsilon, x) = \int_{\epsilon}^{\epsilon+\Delta\epsilon} \chi(\epsilon', x)d\epsilon'. \qquad (11.53)$$

Substituting here Eq. (11.51), we obtain, with scaling the spectrum by $\hbar_{\text{eff}}$, $\epsilon/\hbar_{\text{eff}} \to \epsilon$, to achieve a concise notation,

$$\Delta\chi(\epsilon, x) = \frac{2}{\sqrt{N|x|\ln|x|}} \exp\left[i\left(\epsilon + \frac{\Delta\epsilon}{2}\right)\ln|x|\right] \sin\frac{\Delta\epsilon\ln|x|}{2}. \qquad (11.54)$$

This solution is already square integrable and has the following normalization form

$$I = \lim_{\Delta\epsilon\to 0} \frac{1}{\Delta\epsilon} \int_{-\infty}^{\infty} dx\,|\Delta\chi(\epsilon, x)|^2 = \hbar_{\text{eff}}. \qquad (11.55)$$

Carrying out the following variable change, $z = (\Delta\epsilon/2)\ln|x|$, we represent the integral in Eq. (11.55) as follows,

$$I = \frac{4}{N}\int_{-\infty}^{\infty} \frac{\sin^2 z}{z^2}dz = \frac{2}{N}\Re\int_{-\infty}^{\infty} \frac{1 - e^{2iz}}{z^2}dz. \qquad (11.56)$$

We consider the integral as the principal value of the improper integral

$$I = \text{P.V.}\int_{-\infty}^{\infty} f(z)dz = \lim_{\substack{R\to\infty\\\rho\to 0}} \left[\int_{-R}^{-\rho} \frac{1 - e^{2iz}}{z^2}dz + \int_{\rho}^{\infty} \frac{1 - e^{2iz}}{z^2}dz\right].$$
$$(11.57)$$

The analytical continuation $z = re^{i\theta}$ to the upper half plane yields

$$\int_C f(z)dz = \int_{-\infty}^{-\rho} f(z)dz + \int_{C_\rho} f(z)dz + \int_{\rho}^{\infty} + \int_{C_\infty} f(z)dz$$
$$= 2\pi i \,\text{Res}\,[f(z), z = 0]. \qquad (11.58)$$

Taking into account that $\int_{C_\infty} f(z)dz = 0$, $\text{Res}\,[f(z), z = 0] = -2i$, and $\int_{C_\rho} f(z)dz = 2\pi$, we obtain from Eqs. (11.55), (11.56), and (11.58) that $I = 2\pi$, which yields the normalization constant $N = 4\pi\hbar_{\text{eff}}$.

11.6.2 *Dynamics of the initial wave function*

To study the dynamics of the initial wave function $\Psi_0(x)$, we expand $\Psi_0(x)$ in the complete set of $\chi_\epsilon(x)$:

$$\Psi_0(x) = \int d\epsilon q(\epsilon)\chi_\epsilon(x). \tag{11.59}$$

Note that the explicit form of the expansion coefficients $q(\epsilon)$ is not important, since the integration over the energy ϵ will be performed with exactly the same form of $\chi_\epsilon(x)$. Substituting Eq. (11.59) into the integral in Eq. (11.49), we obtain for the MSD

$$\langle x^2(t)\rangle = 2\int_{-\infty}^{0} dx\, x^2 \int d\epsilon \int d\epsilon'\, q^*(\epsilon')q(\epsilon)\frac{4\epsilon\epsilon'}{\hbar_{\text{eff}}^4}e^{-2it\frac{(\epsilon^2-\epsilon'^2)}{\hbar_{\text{eff}}^3}}\chi_{\epsilon'}^*(x)\chi_\epsilon(x), \tag{11.60}$$

where we use that $\chi_\epsilon(-x)=\chi_\epsilon(x)$. The complex Gaussian exponents are represented in the form of the Fourier integrals,

$$e^{\pm 2it(\epsilon^2/\hbar_{\text{eff}}^3)} = \int_{-\infty}^{\infty}\frac{e^{\mp i\tau\epsilon}d\tau}{\sqrt{\pm 8\pi it/\hbar_{\text{eff}}^3}}\exp\left[\mp\frac{i\hbar_{\text{eff}}^3\tau^2}{8t}\right]. \tag{11.61}$$

Substituting these expressions into Eq. (11.60) and taking into account the explicit form of $\chi_\epsilon(x)$, we obtain

$$\epsilon(\ln|x|-\hbar_{\text{eff}}\tau)/\hbar_{\text{eff}} = \epsilon\ln(|x|e^{-\hbar_{\text{eff}}\tau})/\hbar_{\text{eff}}. \tag{11.62}$$

Then the term $4\epsilon\epsilon'/\hbar_{\text{eff}}^4$ is replaced by the second derivative $\frac{4}{\hbar_{\text{eff}}^4}\frac{d^2}{d\tau d\tau'}$, and we integrate with respect to ϵ and ϵ' to obtain the following expression,

$$\begin{aligned}
\langle x^2(t)\rangle &= 2\int_0^{\infty}dx\,x^2\int d\tau\int d\tau' e^{-i\frac{\hbar_{\text{eff}}^3(\tau^2-\tau'^2)}{8t}}\frac{4}{\hbar_{\text{eff}}^4}\frac{d^2}{d\tau d\tau'}\\
&\quad \times e^{\frac{\hbar_{\text{eff}}(\tau+\tau')}{2}}\Psi_0^*\left(xe^{-\hbar_{\text{eff}}\tau'}\right)\Psi_0\left(xe^{-\hbar_{\text{eff}}\tau}\right)\frac{1}{8\pi t/\hbar_{\text{eff}}^3}\\
&= \int_0^{\infty}dx\,x^2\int d\tau\int d\tau' e^{-i\frac{\hbar_{\text{eff}}^3(\tau^2-\tau'^2)}{8t}}e^{\frac{\hbar_{\text{eff}}(\tau+\tau')}{2}}\frac{1}{8\pi t/\hbar_{\text{eff}}}\\
&\quad \times\left[\Psi_0^*\left(xe^{-\hbar_{\text{eff}}\tau'}\right)\Psi_0\left(xe^{-\hbar_{\text{eff}}\tau}\right)-2x\Psi_0^{*\prime}\left(xe^{-\hbar_{\text{eff}}\tau'}\right)\Psi_0\left(xe^{-\hbar_{\text{eff}}\tau}\right)\right.\\
&\quad -2x\frac{\hbar_{\text{eff}}}{2}\Psi_0^*\left(xe^{-\hbar_{\text{eff}}\tau'}\right)\Psi_0'\left(xe^{-\hbar_{\text{eff}}\tau}\right)\\
&\quad \left.+4x^2\Psi_0^{*\prime}\left(xe^{-\hbar_{\text{eff}}\tau'}\right)\Psi_0'\left(xe^{-\hbar_{\text{eff}}\tau}\right)\right], \tag{11.63}
\end{aligned}$$

where the prime on the wave function denotes differentiation with respect to the argument. The next step is integration with respect to τ and τ'. To this end, we perform the variables changes $\tau=(u+v)/\hbar_{\text{eff}}$ and $\tau'=$

$(u - v)/\hbar_{\text{eff}}$, with the Jacobian of the transformation equal to $2/\hbar_{\text{eff}}^2$. We define $y = xe^{-\hbar_{\text{eff}}(\tau+\tau')/2}$. The integration with respect to u is exact and yields the δ function, $\delta(v + \nu i t/\hbar_{\text{eff}})$, where $\nu = 4, 5, 6$. Therefore the integration with respect to v is also exact. Finally, the expectation value reads

$$\langle x^2(t) \rangle = \int_0^\infty \left[y^2 \left| \Psi_0 \left(y e^{4it/\hbar_{\text{eff}}} \right) \right|^2 - 2y^3 \left(\left| \Psi_0 \left(y e^{5it/\hbar_{\text{eff}}} \right) \right|^2 \right)' \right. $$
$$\left. + 4y^4 \left(\left| \Psi_0 \left(y e^{6it/\hbar_{\text{eff}}} \right) \right|^2 \right)'' \right] dy. \qquad (11.64)$$

11.6.3 *Exponential spreading and singular behavior of quantum dynamics*

Our result has the remarkable feature that any "good" initial wave functions, such as Gaussian or exponential functions, lead to singular behavior for the expectation values of physical operators. For example, let us consider the initial wave function $\Psi_0(x)$ as a simple exponential,

$$\Psi_0(x) = \frac{\sqrt{a} e^{-a|x|/2}}{\sqrt{2}}, \qquad (11.65)$$

which is square normalized. However, its dynamics according to the result in Eq. (11.64) leads to singular behavior. As an example, let us estimate the first term of the MSD in Eq. (11.64),

$$\int_0^\infty dy\, y^2 \left| \Psi_0 \left(y e^{4it/\hbar_{\text{eff}}} \right) \right|^2 = \frac{a}{2} \int_0^\infty dx\, x^2 \exp[-ax \cos(4t/\hbar_{\text{eff}})]. \qquad (11.66)$$

Therefore, an "explosion" occurs in the system at the time $t = \hbar_{\text{eff}} \pi/8$ with the MSD $\langle x^2(t) \rangle = \infty$. Note that this result is a pure quantum effect and does not have any counterpart in the classical limit, since the nonlinearity parameter here is $\mu = 1/\hbar_{\text{eff}}^2$. Therefore, the explosion time can be written as $\pi/8\mu\hbar_{\text{eff}}$, which corresponds to the results obtained in Refs. [184, 185]. This result implies that the choice of the initial condition needs special care.

Obviously these singularities result from a specific preparation of the initial wave packets $\Psi_0(x)$. Let us choose the initial conditions "properly" to obtain finite moments of the physical variables. In light of Eq. (11.59), we represent the initial condition as the spectral decomposition with the Gaussian weight $q(\epsilon) = \left[\frac{2a}{\pi} \right]^{\frac{1}{4}} \exp(-a\epsilon^2)$, where a is real and positive. This yields an initial wave packet in the form of a log-normal distribution,

$$\Psi_0(x) = \frac{1}{\sqrt{N}} \exp \left(-\frac{1}{4a\hbar_{\text{eff}}^2} \ln^2 |x| - \frac{1}{2} \ln|x| \right), \qquad (11.67)$$

with $N = 4a\hbar_{\text{eff}}\sqrt{\pi}$. Using Eqs. (11.67) and (11.64), we obtain for the n-th moment of $\hat{x}$,

$$\langle \hat{x}^n(t) \rangle = \frac{1}{2a\hbar_{\text{eff}}\sqrt{\pi}} \exp\left(\frac{\nu^2 t^2}{4a\hbar_{\text{eff}}^4}\right)$$

$$\times \int_0^\infty \exp\left(-\frac{1}{2a\hbar_{\text{eff}}^2}\ln^2 x + n\ln x\right) d(\ln x)$$

$$= \frac{1}{\sqrt{2a}}\exp\left(\frac{a\hbar_{\text{eff}}^2 n^2}{2} + \frac{\nu^2 t^2}{4a\hbar_{\text{eff}}^4}\right). \tag{11.68}$$

The expectation values are finite and grow exponentially for arbitrary long times. This exponential growth with time is universal for all n-th moments. From a mathematical point of view, this exponential growth relates to the action of the evolution operator $e^{-ibx\hbar_{\text{eff}}\frac{d}{dx}}$, where[8]

$$e^{ibx\frac{d}{dx}}f(x) = f\left(xe^{ib}\right). \tag{11.69}$$

This term is due to the quantum dynamics near the hyperbolic point. However, its nonlinear effect is due to the comb geometry.

Our discussion of the comb model shows that the fractional time derivative in quantum mechanics does have physical relevance. The destruction of the quantum unitary transformation by the fractional time derivative is associated with the interaction of the Hamiltonian system with the environment. Another important point is that the fractional derivative results in nonlinear quantum dynamics along the backbone.

---

[8] See, for example, [184].

Chapter 12

Path integral description of combs

12.1 Fractional Brownian motion and path integral formalism

We start from an interesting result by Sebastian [187] on the path integral of a free particle. His consideration does not relate to the FFPE, but it is instructive to consider this result for two reasons. First, it is a beautiful exercise in fractional calculus. Second, it provides a deep insight into the fractional path integral and an illustration of fractional Brownian motion. Fractional Brownian motions (fBm) $B_H(t)$ [160, 188] are a family of Gaussian processes characterized by a parameter H, $0 < H < 1$, whose correlation function is given by

$$\langle B_H(t)B_H(s) \rangle = \frac{1}{2} \left(t^{2H} + s^{2H} - |t - s|^{2H} \right), \tag{12.1}$$

where H is the so-called Hurst parameter or Hurst index.

In one dimension, let us consider the random process $X(t)$ that describes the position of a particle at time t having started at X_0 at $t = 0$ [187],

$$X(t) - X_0 = \mathcal{D}^{-\alpha} \xi(t) = \frac{1}{\Gamma(\alpha)} \int_0^t (t - t')^{\alpha - 1} \xi(t') dt', \tag{12.2}$$

where $\alpha \in (0, 1)$ and $\xi(t)$ is the Gaussian white noise defined in Eq. (5.3), $\langle \xi(t)\xi(t') \rangle = 2D\delta(t - t')$. For $1/2 < \alpha < 3/2$, this random process coincides with Paul Lévy's definition of fBm [160] where $\alpha = H + 1/2$. Following Ref. [187], we construct the path integral representation of this motion. Adopting the notation $\delta X(t) = X(t) - X_0$, Eq. (12.2) can be rewritten as follows

$$\mathcal{D}^\alpha \delta X(t) = \xi(t). \tag{12.3}$$

Here we recall the property

$$1 \equiv \mathcal{D}^\alpha \mathcal{D}^{-\alpha} = \mathcal{D}\mathcal{D}^{\alpha-1}\mathcal{D}^{-\alpha} = \mathcal{D}\mathcal{D}^{-1}, \tag{12.4}$$

which is valid for $0 < \alpha < 1$. From the definition (5.10) of the pdf

$$P_1(x,t) = \int \delta(X(t) - x)[d\rho(\xi)], \qquad (12.5)$$

we obtain the conditional probability of finding the particle at the position $X(t)$ at time t, when it was at X_0 at $t = 0$. Following Sebastian's arguments that the measure in the path integral representation for the white noise $\xi(t)$ is Gaussian and $\xi(t)$ is linearly related to $X(t)$, the measure for the latter is also Gaussian with respect to relation (12.3). This yields the conditional probability as follows,

$$G\left(X(t), t \mid X_0, 0\right) = \int_{X_0, t=0}^{X(t), t} \exp\left[-\frac{1}{4D}\int_0^t [\mathcal{D}^\alpha \delta X(\tau)]^2 \, d\tau\right] [dX(\tau)].$$
$$(12.6)$$

Let us obtain this propagator by performing the Gaussian functional integration. As the first step of the procedure, we represent the path $\delta X(t)$ as the "extremum path" $\delta X_e(t)$ and the deviation $\delta x(t)$,

$$\delta X(t) = \delta X_e(t) + \delta x(t), \qquad (12.7)$$

where the extremum path is determined by the extremum of the action,

$$S = \frac{1}{4D}\int_0^t [\mathcal{D}^\alpha \delta X(\tau)]^2 \, d\tau. \qquad (12.8)$$

It is also satisfies the boundary conditions

$$\delta X_e(\tau = 0) = 0 \quad \text{and} \quad \delta X(\tau = t) = X(t) - X_0. \qquad (12.9)$$

Consequently, the deviation path has to obey $\delta x(0) = \delta x(t) = 0$. The extremum action corresponds to the first variation of the action being zero, $\delta S = 0$. The first variation reads

$$\delta S = \int_0^t \mathcal{D}^\alpha \delta X_e(\tau) \mathcal{D}^\alpha \delta x(\tau) d\tau. \qquad (12.10)$$

Using the definition for the Riemann–Liouville fractional derivative, see Chap. 2, $\mathcal{D}^\alpha = D\mathcal{D}^{\alpha-1}$ for $0 < \alpha < 1$, we integrate (12.10) by part,

$$\delta S = -\int_0^t \left(\mathcal{D}^{\alpha-1}\delta x(\tau)\right) \mathcal{D}^{\alpha+1}\delta X_e(\tau) d\tau$$
$$+ \mathcal{D}^\alpha \delta X_e(t)\frac{1}{\Gamma(1-\alpha)}\int_0^t (t - \tau)^{-\alpha}\delta x(\tau) d\tau. \qquad (12.11)$$

Here we also take into account that $\mathcal{D}^{\alpha-1}\delta x(\tau)|_{\tau=0} = 0$. Using the Dirichlet's formula (2.72), Eq. (12.11) reads

$$\delta S = -\int_0^t \delta x(\tau)d\tau \int_\tau^t \frac{(\tau-s)^{-\alpha}}{\Gamma(1-\alpha)} \mathcal{D}^{\alpha+1}\delta X_e(s)ds$$
$$+ \mathcal{D}^\alpha \delta X_e(t) \int_0^t \frac{(t-\tau)^{-\alpha}}{\Gamma(1-\alpha)} \delta x(\tau)d\tau . \qquad (12.12)$$

For the extremum path, δS vanishes for arbitrary $\delta x(\tau)$, which yields the following integral equation,

$$_t\mathcal{D}_\tau^{\alpha-1}\mathcal{D}^{\alpha+1}\delta X_e(\tau) = \mathcal{D}^\alpha \delta X_e(t)\frac{(t-\tau)^{-\alpha}}{\Gamma(1-\alpha)}. \qquad (12.13)$$

We recall that $_t\mathcal{D}_\tau^{\alpha-1} \equiv {}^{RL}_t\mathcal{D}_\tau^{\alpha-1}$ is the right Riemann–Liouville fractional derivative.

12.1.1 The extremum path

To solve the integral equation (12.13), we operate on both sides with $_t\mathcal{D}_\tau^{-\alpha}$. Then on the left side of the equation, we obtain $_t\mathcal{D}_\tau^{-\alpha}{}_t\mathcal{D}_\tau^{\alpha-1} = {}_t\mathcal{D}_\tau^{-1}$, while on the right side the fractional integration yields $_t\mathcal{D}_\tau^{-\alpha}(t-\tau)^{-\alpha} = \Gamma(1-\alpha)$. The latter is defined up to a function that vanishes being acted upon with the fractional derivative.[1] Therefore we add the function $-C(t-\tau)^{\alpha-1}$, where the constant C is defined by the boundary conditions. After the fractional integration, Eq. (12.13) reads

$$\int_\tau^t \mathcal{D}^{\alpha+1}\delta X_e(\tau)d\tau = \int_\tau^t \frac{d^2}{d\tau^2}\int_0^\tau \frac{(\tau-s)^{-\alpha}}{\Gamma(\alpha)} \delta X_e(s)dsd\tau$$
$$= \mathcal{D}^\alpha \delta X_e(t) - C(t-\tau)^{\alpha-1}. \qquad (12.14)$$

Integration with respect to τ results in

$$\mathcal{D}^\alpha \delta X_e(\tau) = C(t-\tau)^{\alpha-1}. \qquad (12.15)$$

To find the constant C, we have to solve the inhomogeneous integral equation (12.15). Applying the integral operator $\mathcal{D}^{-\alpha}$ to the both sides of Eq. (12.15), we have the following sequence of operations for the left side of the equation,

$$\mathcal{D}^{-\alpha}\mathcal{D}^\alpha \delta X_e(\tau) = \mathcal{D}^{-\alpha}\mathcal{D}\mathcal{D}^{\alpha-1}\delta X_e(\tau)$$
$$\equiv \mathcal{D}^{-\alpha}\mathcal{D}F(\tau) = \mathcal{D}^{1-\alpha}F(\tau) - \frac{F(0)}{\Gamma(\alpha)}\tau^{\alpha-1}, \qquad (12.16)$$
$$\mathcal{D}^{1-\alpha}\mathcal{D}^{\alpha-1}\delta X_e(\tau) - C_1\tau^{\alpha-1} = \delta X_e(\tau) - C_1\tau^{\alpha-1}. \qquad (12.17)$$

---

[1]This corresponds to the integral equation $_t\mathcal{D}_\tau^\alpha f(\tau) = 0$, which has the solution $f = (t-\tau)^{\alpha-1}$.

Here we used the notation $F(\tau) = \mathcal{D}^{\alpha-1}\delta X_e(\tau)$ and Eq. (2.75), while the constant $C_1 = \frac{F(0)}{\Gamma(\alpha)}$ will be determined from the boundary conditions. We obtain the evolution of the extremum path in explicit form,

$$\delta X_e(\tau) = \frac{C}{\Gamma(\alpha)} \int_0^\tau d\tau'(\tau - \tau')^{\alpha-1}(t - \tau')^{\alpha-1} + C_1\tau^{\alpha-1}. \qquad (12.18)$$

From the boundary conditions in Eq. (12.9), we obtain $C_1 = 0$. The constant C is obtained from the integration, which yields

$$C = (X(t) - X_0)\Gamma(\alpha)(2\alpha - 1)t^{1-2\alpha}. \qquad (12.19)$$

Taking into account Eq. (12.19), we obtain

$$\mathcal{D}^\alpha \delta X_e(\tau) = (X(t) - X_0)\Gamma(\alpha)(2\alpha - 1)(t - \tau)^{\alpha-1}t^{2\alpha-1}, \qquad (12.20)$$

which yields the action (12.8) of the extremum path,

$$S_e = \frac{(2\alpha - 1)\Gamma^2(\alpha)}{4D} \frac{(X(t) - X_0)^2}{t^{2\alpha-1}}. \qquad (12.21)$$

We can finally obtain an explicit expression for the propagator (12.6),

$$G\left(X(t), t \mid X_0, 0\right) = e^{-S_e} \int_{0,t=0}^{0,t} \exp\left[-\frac{1}{4D}\int_0^t \left[\mathcal{D}^\alpha \delta x(\tau)\right]^2 d\tau\right] [d\delta x(\tau)]$$

$$= \exp\left[\frac{(2\alpha - 1)\Gamma^2(\alpha)}{4D} \frac{(X(t) - X_0)^2}{t^{2\alpha-1}}\right] G\left(0, t \mid 0, 0\right). \qquad (12.22)$$

The deviation path propagator $G\left(0, t \mid 0, 0\right)$ is easily found from the normalization condition,

$$\int_{-\infty}^\infty G\left(X(t), t \mid X_0, 0\right) dX(t) = 1, \qquad (12.23)$$

which yields

$$G\left(X(t), t \mid X_0, 0\right) = \left[\frac{(2\alpha - 1)\Gamma^2(\alpha)}{4Dt^{2\alpha-1}\pi}\right]^{\frac{1}{2}} \exp\left[\frac{(2\alpha - 1)\Gamma^2(\alpha)}{4D} \frac{(X(t) - X_0)^2}{t^{2\alpha-1}}\right]. \qquad (12.24)$$

This expression has been also obtained in an alternative way in Ref. [189], and the path integral approach is reviewed in [190].

To conclude this section, we infer the relation, which is not obvious, between the pdf (12.5) and the conditional probability (12.6).

Equation (12.5) can be rewritten by means of the chain rule

$$P_1(x, t) = \int_{-\infty}^\infty G(x, t \mid x_0, 0) P_0(x_0) dx_0. \qquad (12.25)$$

Partitioning the time interval $[0, t]$ into N subintervals with $N \to \infty$, we obtain from Eq. (12.25) a product of Green's functions, or conditional probabilities, each of which is defined on the infinitesimal time scale $dt = \lim_{N \to \infty} \frac{t}{N}$. The solution of the Langevin equation (12.3) on the subintervals is

$$dt \mathcal{D}^{\alpha} \delta X(t) = \xi(t)dt, \qquad (12.26)$$

which is accounted for by the Green's function

$$G(x, t | x_0, 0) = \prod_{\tau=0}^{t} \frac{1}{d\tau} \delta\left(\mathcal{D}^{\alpha}(\delta X(\tau)) - \xi(\tau) \right), \qquad (12.27)$$

where $\delta X(\tau) = X(\tau + d\tau) - X(\tau)$. Substituting this expression into Eq. (12.5), we obtain the conditional probability (12.6).

12.2 Path integral of the comb

Let us return to the formal definition of the pdf in Eq. (5.10)

$$P(x, y, t) = \langle\langle \delta[X(t) - x] \delta[Y(t) - y] \rangle\rangle, \qquad (12.28)$$

and obtain the Green's function in the form of a path integral. The functionals (5.8) and (5.9) for the random trajectories imply that Eq. (12.28) can be rewritten for the transition probability as follows,

$$
\begin{aligned}
G(x, y, t | x', y', t') \\
= \int \delta\left(x - x' - \int_{t'}^{t} C[y(\tau)] \xi_x(\tau) d\tau \right) e^{-\frac{1}{4} \int_{t'}^{t} \xi_x^2(\tau) d\tau} [d\xi_x(\tau)] \\
\times \int \delta\left(y - y' - \int_{t'}^{t} \xi_y(\tau) d\tau \right) e^{-\frac{1}{4} \int_{0}^{t} \xi_y^2(\tau) d\tau} [d\xi_y(\tau)], \qquad (12.29)
\end{aligned}
$$

where we set[2] $D_x = D_y = 1$, and we take $t' = 0$ with $x' = x_0$ and $y' = y_0$ as the initial condition. Integration with respect to $\xi_y(\tau)$ with the δ-function is straightforward and yields the Wiener process,

$$
\begin{aligned}
\int \delta\left(y - y_0 - \int_{0}^{t} \xi_y(\tau) d\tau \right) e^{-\frac{1}{4} \int_{0}^{t} \xi_y^2(\tau) d\tau} [d\xi_y(\tau)] \\
= \int e^{-\frac{1}{4} \int_{0}^{t} \dot{y}^2(\tau) d\tau} [dy(\tau)]. \qquad (12.30)
\end{aligned}
$$

---

[2] Instead we can work with dimensionless variables and parameters, where the condition $D_x = D_y = 1$ is fulfilled.

Here the kinetic energy corresponds to the Lagrangian $m\dot{y}^2/2$ with $m = 1/2$. To integrate with respect to $\xi_x(\tau)$, we first represent the δ-function in the form of the integral

$$\delta\left(x - x_0 - \int_0^t C[y(\tau)]\xi_x(\tau)d\tau\right)$$
$$= \frac{1}{2\pi}\int_{-\infty}^{\infty}\exp\left[-iu\left(x - x_0 - \int_{t'}^t C[y(\tau)]\xi_x(\tau)d\tau\right)\right]. \qquad (12.31)$$

Therefore, integrating with respect to $\xi_x(\tau)$ we obtain

$$\pi^{-\frac{1}{2}}\exp\left[-u^2\int_0^t C^2[y(\tau)]d\tau\right] = \pi^{-\frac{1}{2}}\exp\left[-u^2\int_0^t \delta(y)d\tau\right], \qquad (12.32)$$

where we took into account that for the comb geometry $C^2(y) = \delta(y)$, see page 84. Now the Green's function reads

$$G(x, y, t|x_0, y_0, 0) = \frac{1}{\sqrt{\pi}}\int_{-\infty}^{\infty}due^{-iu(x-x_0)}\int_{y(0)=y_0}^{y(t)=y}e^{-\int_0^t L(\dot{y},y;u^2)d\tau}[dy(\tau)]. \qquad (12.33)$$

Here, the Lagrangian

$$L(\dot{y}, y) \equiv L(\dot{y}, y; u^2) = \tfrac{1}{4}\dot{y}^2 + u^2\delta(y) \qquad (12.34)$$

describes the quantum dynamics of a particle with a half unit mass in the δ potential. This model has attracted considerable attention, and different aspects of the problem have been studied in a series of publications[3] [180,191–193]. It also corresponds to the Green's function considered in Chapter 11 by replacing $-1 \to i$ and $\lambda = -u^2$.

12.2.1 *Dirac delta potential*

Let us calculate the functional integral for the y-Green's function,

$$G_y(T) \equiv G_y(y, T | y_0, 0) = \int_{y(0)=y_0}^{y(T)=y}e^{-\int_0^t L(\dot{y},y;\lambda)d\tau}[dy(t)]. \qquad (12.35)$$

Following Ref. [191], we represent the action as a sum of two actions $S = \int_0^t L(\dot{y}, y)dt = S_0 + S_1$, where S_0 corresponds to a Wiener process, while $S_1 = -\lambda\int_0^T \delta(y)dt$ corresponds to the Dirac delta potential and $\lambda = -u^2$. Therefore $G_y(T)$ is represented as the expansion

$$G_y(T) = \int e^{-S_0}\left(1 + \sum_{n=1}^{\infty}\frac{1}{n!}S_1^n\right)[dy(t)] = G_y^0(T) + G_y^1(T), \qquad (12.36)$$

---

[3] See the Supplements in the reprint of Ref. [181] in 2005.

where $G_y^0(T) = \sqrt{\frac{1}{4\pi T}}e^{-\frac{(y-y_0)^2}{4T}}$ is the free particle propagator. Our main concern is the potential term,

$$G_y^1(T) = \sum_{n=1}^{\infty} G_n(y, T \mid y_0, 0), \quad \text{where } G_n(y, T \mid y_0, 0) = \frac{1}{n!}\int e^{-S_0} S_1^n [dy(t)].$$
(12.37)

Note that the path integration supposes time ordering, which yields

$$S_1^n = (-\lambda)^n \int_0^T dt_n \int_0^T dt_{n-1}\cdots \int_0^T dt_1 \prod_{k=1}^n \delta\left(y(t_k)\right)$$

$$= (-\lambda)^n n! \int_0^T dt_n \int_0^{t_n} dt_{n-1}\cdots \int_0^{t_2} dt_1 \prod_{k=1}^n \delta(y_k). \quad (12.38)$$

As the time partition we use $t_0 = 0$, $t_{n+1} = T$, and $t_0 < t_1 < \cdots < t_n < T$.
We employ the chain rule for $G_y^0(T)$, which reads

$$G_y^0(T) \equiv G_y^0(y, T \mid y_0, 0) = \int_{-\infty}^{\infty} dy_n\, G_y^0(y, T \mid y_n, t_n) G_y^0(y_n, t_n \mid y_0, 0)$$

$$= \int_{-\infty}^{\infty} dy_n \cdots \int_{-\infty}^{\infty} dy_1\, G_y^0(y, T \mid y_n, t_n) G_y^0(y_n, t_n \mid y_{n-1}, t_{n-1})$$

$$\cdots G_y^0(y_1, t_1 \mid y_0, 0). \quad (12.39)$$

Therefore we have

$G_n(y, T \mid y_0, 0)$

$$= (-\lambda)^n \int_0^T dt_n \int_0^{t_n} dt_{n-1}\cdots \int_0^{t_2} dt_1$$

$$\times \int_{-\infty}^{\infty} dy_n \cdots \int_{-\infty}^{\infty} dy_1\, G_y^0(y, T \mid y_n, t_n)\cdots G_y^0(y_1, t_1 \mid y_0, 0) \prod_{k=1}^n \delta(y_k)$$

$$= (-\lambda)^n \int_0^T dt_n \int_0^{t_n} dt_{n-1}\cdots \int_0^{t_2} dt_1 G_y^0(y, T \mid 0, t_n)$$

$$\times \prod_{k=2}^n G_y^0(0, t_k \mid 0, t_{k-1}) G_y^0(0, t_1 \mid y_0, 0). \quad (12.40)$$

This chain of the conditional probabilities, or of Green's functions, is suitable for the Laplace transformation, which yields simply a product of the Laplace transforms,

$$\hat{G}_n(y, s \mid y_0, 0) = (-\lambda)^n \hat{G}_y^0(y, s)\hat{G}_y^0(y_0, s)\left[\hat{G}_y^0(0, s)\right]^{n-1}, \quad (12.41)$$

where the Laplace transforms of the Wiener processes[4] and loops (or traps) are, respectively,

$$\hat{G}_y^0(y, s)\hat{G}_y^0(y_0, s) = \frac{1}{4s}e^{-\sqrt{s}(|y|+|y_0|)}, \quad (12.42a)$$

---

[4]We take into account the result of the integration (11.26).

$$\hat{G}_y^0(0,s) = \frac{1}{2\sqrt{s}}. \tag{12.42b}$$

Eventually we obtain for the potential part of the Green's function,

$$G_y^1(T) \equiv G_y^1(y, T \mid y_0, 0) = \mathcal{L}^{-1}\left[\sum_{n=1}^{\infty}(-\lambda)^n \left(\frac{1}{2\sqrt{s}}\right)^{n+1} e^{-\sqrt{s}(|y|+|y_0|)}\right]$$

$$= \frac{-\lambda}{2} \cdot \mathcal{L}^{-1}\left[\frac{e^{-\sqrt{s}(|y|+|y_0|)}}{\sqrt{s}\left(\sqrt{s}+\lambda/2\right)}\right]. \tag{12.43}$$

This result has been studied in Chap. 11.

Performing the inverse Laplace transformation in Eq. (12.43), we have the integral

$$G_y^1(y, T \mid y_0, 0) = \frac{-\lambda}{4\pi i} \int_{c-i\infty}^{c+i\infty} e^{sT} \frac{e^{-\sqrt{s}(|y|+|y_0|)}}{\sqrt{s}\left(\sqrt{s}+\lambda/2\right)} ds. \tag{12.44}$$

Using the same arguments as for the integral in Eq. (11.20) with $\sqrt{s} = z$, we arrive at the Gaussian integration, which yields

$$G_y^1(y, T \mid y_0, 0) = \frac{-\lambda}{2\sqrt{4\pi T}} \int_0^{\infty} e^{-(|y|+|y_0|+\zeta)^2/4T} e^{-\zeta\lambda/2} d\zeta. \tag{12.45}$$

Finally the Green's function found here,

$$G_y(T) = G_y^0(y, T \mid y_0, 0) + \frac{-\lambda^2}{2} \int_0^{\infty} G_y^0(|y|+|y_0|+\zeta, T \mid 0, 0) e^{-\zeta\lambda/2} d\zeta, \tag{12.46}$$

agrees exactly with the result obtained in an alternative way by Gaveau and Schulman [180].

12.3   Scattering considerations: Green's function

It is instructive to present here some tutorial arguments[5] on the scattering Green's function [194,195]. As discussed in the previous chapter, the form of the solution of Eq. (11.14) depends on the form of the boundary conditions. The dimensionless counterpart of the eigenvalue problem, including both Schrödinger and Fokker–Planck equations, reads[6]

$$\left[-\frac{d^2}{dy^2} - \lambda\delta(y)\right]\Psi_e(y) = e\Psi_e(y). \tag{12.47}$$

---

[5]We follow the presentation by Blinder [192] on scattering and the Dirac δ-function potential.

[6]It corresponds to the Hamiltonian $\frac{p^2}{2m} - \lambda\delta(y)$ with the mass $m = 1/2$, and the classical momentum is replaced by its quantum counterpart $p \to -i\frac{d}{dy}$.

For the Sturm–Liouville or boundary value problem (12.47), we assume free boundary conditions at infinity. Considering the $\delta(y)$-potential as an attractive potential, one obtains two classes of the solutions. The first class corresponds to the bound states and contains only one state,

$$\Psi_0(y) = \sqrt{\lambda/2}e^{-\lambda|y|/2}, \quad e_0 = -\lambda^2/4. \tag{12.48}$$

Note that in contrast to the previous chapter, here $\Psi_0(y)$ denotes the ground state wave function.

The second class corresponds to the scattering problem and contains the continuous spectrum solutions, which are even and odd standing waves, according the symmetry properties of the equation,

$$\Psi_k^{\text{odd}}(y) = \pi^{-1/2}\sin(ky), \tag{12.49a}$$

$$\Psi_k^{\text{even}}(y) = [\pi(4k^2 + \lambda^2)]^{-1/2} \cdot [2k\cos(ky) - \lambda\sin(k|y|)], \tag{12.49b}$$

where $e_k = k^2$ and $k > 0$.

By virtue of the solutions (12.49), we derive the Green's function by means of its standard definition in energy space, see e.g., Ref. [194]. In our case it corresponds to the Laplace transform

$$\hat{G}(y, y_0, s) = \sum_k \frac{\Psi_k^*(y_0)\Psi_k(y)}{s - e_k} = \frac{\Psi_0(y)\Psi_0(y_0)}{s + \lambda^2/4} + \int_0^\infty \frac{\Psi_k^*(y_0)\Psi_k(y)}{s - k^2}dk. \tag{12.50}$$

The singularities of the integrand are simple poles, and the integration is a standard calculation of residues, which yields the expression in Eq. (12.43), namely,

$$\hat{G}(y, y_0, s) = \frac{1}{\sqrt{s}}e^{-\sqrt{s}|y-y_0|} - \frac{\lambda}{2}\frac{e^{-\sqrt{s}(|y|+|y_0|)}}{\sqrt{s}(\sqrt{s}+\lambda/2)}. \tag{12.51}$$

Solutions (12.49) can also be considered as the result of the scattering by the δ potential of left and right incident waves [192],

$$\Psi_k^\pm(y) = \frac{1}{\sqrt{2\pi}}\left[e^{\pm iky} + \mathcal{B}(k)e^{ik|y|}\right], \tag{12.52}$$

where $\mathcal{B}(k) = -\lambda/(\lambda + 2ik)$ is the so called scattering amplitude, determined by Eq. (12.47).

The Green's function (12.50) can alternatively be obtained from Sturm–Liouville theory, when the solution to the equation[7]

$$\left[k^2 + \frac{d^2}{dy^2} + \lambda\delta(y)\right]\Psi_k(y) = \delta(y - y_0) \tag{12.53}$$

---

[7]The function $G(x, y)$ is called the influence function or Green's function of the boundary-value problem described by inhomogeneous equation $\frac{d}{dx}(p(x)g'(x) + q(x)g(x) = r(x, y)$, with boundary conditions $g(x_1) = g_1$ and $g(x_2) = g_2$, and where the Dirac delta function is a particular case of $r(x, y) = \delta(x - y)$. In our case $p = 1$.

can be obtained from the properties of the Green's function at the boundaries [196]. This yields in k-space (energy space),

$$G(y, y_0, k) = \frac{\Psi_k^*(y_0)\Psi_k(y)}{pW(y_0)} = \frac{\Psi_k^*(y_0)\Psi_k(y)}{\Psi_k^*(y_0)\Psi_k'(y_0) - \Psi_k^{*'}(y_0)\Psi_k(y_0)}, \qquad (12.54)$$

where $W(y_0)$ is the Wronskian, $W(x) = W(z_1(x), z_2(x)) = z_1(x)z_2'(x) - z_2(x)z_1'(x)$ and $p = 1$ (see the Footnote 7). Accounting for the boundary conditions of the scattering problem for the case $0 < y_0 < y$, we use $\Psi_k(y) = e^{iky}$, $\Psi_k^*(y_0) = e^{-iky_0} + \mathcal{B}(k)e^{iky_0}$, which yields

$$G(y, y_0, k) = (ik)^{-1}e^{iky}\left[e^{-iky_0} + \mathcal{B}(k)e^{iky_0}\right]. \qquad (12.55)$$

Taking into account that the solution is symmetric with respect to $0 > y_0 > y$, we have

$$G(y, y_0, k) = (ik)^{-1}\left[e^{ik|y-y_0|} + \mathcal{B}(k)e^{ik(|y|+|y_0|)}\right]. \qquad (12.56)$$

Setting $ik = -\sqrt{s}$, which is a transition from the inverse Laplace transformation to the inverse Fourier transformation, we arrive at Eq. (12.51).

12.4 Scattering considerations: Supersymmetric quantum mechanics

In the following we adopt the standard supersymmetric quantum theory[8] to deal with the boundary value problem in Eq. (12.47) and write it as an eigenvalue problem with the Hamiltonian (Schrödinger operator)

$$\mathcal{H}^{(1)} = -\frac{d^2}{dy^2} + V^{(1)}(y), \qquad (12.57)$$

where $V^{(1)} = -\lambda\delta(y)$. As discussed above, the discrete spectrum contains only one state, the ground state $\Psi_0(y)$ with energy $e_0 = -\lambda^2/4$, see Eq. (12.48). To write the Hamiltonian (12.57) in multiplicative form, we write the ground state in the following form,

$$\Psi_0(y) = \exp\left[-F_W(y)\right], \quad \text{where} \quad F_W(y) = \int^y W(y)dy. \qquad (12.58)$$

This allows us to factorize the Hamiltonian as follows,

$$\mathcal{H}^{(1)} - e_0 = \mathcal{A}^\dagger\mathcal{A}, \qquad (12.59)$$

where the creation and annihilation operators, $\mathcal{A}^\dagger$ and $\mathcal{A}$, are given by

$$\mathcal{A} = \frac{d}{dy} + W(y) \equiv \frac{d}{dy} + F_W' \equiv \Psi_0\frac{d}{dy}\Psi_0^{-1}, \qquad (12.60a)$$

---

[8]For a comprehensive discussion of quantum mechanics and supersymmetry see, e.g., Ref. [197].

$$\mathcal{A}^\dagger = -\frac{d}{dy} + W(y) \equiv -\frac{d}{dy} + F'_W \equiv -\Psi_0 \frac{d}{dy} \Psi_0^{-1}. \qquad (12.60b)$$

This procedure is based on the so-called Darboux–Crum transformation[9] [199]. The quantity $W(y)$ is generally referred to as the "superpotential" and can be written in terms of the ground state wave function, $W(y) = -\Psi'_0(y)/\Psi_0(y)$. One can also construct the supersymmetric partner Hamiltonian,

$$\mathcal{H}^{(2)} = \mathcal{A}\mathcal{A}^\dagger + e_0 = -\frac{d^2}{dy^2} + V^{(2)}(y). \qquad (12.61)$$

The supersymmetric partner potentials $V_{1,2}(y)$ are determined by the superpotential,

$$V^{(1)}(y) = e_0 + W^2(y) - W'(y), \qquad (12.62a)$$
$$V^{(2)}(y) = e_0 + W^2(y) + W'(y). \qquad (12.62b)$$

Example 12.1. We employ expressions (12.62) explicitly to obtain the partner potentials. From Eq. (12.58), we have $F_W(y) = -\ln(\Psi_0(y))$ and

$$W(y) = F'_W(y) = -\frac{\Psi'_0(y)}{\Psi_0(y)} = \frac{\lambda}{2}[2\theta(y) - 1]. \qquad (12.63)$$

Therefore $W'(y) = \lambda\delta(y)$ and $W^2(y) = \frac{\lambda^2}{4}$, and eventually we obtain

$$V^{(1)} = -\lambda\delta(y), \quad V^{(2)} = \lambda\delta(y). \qquad (12.64)$$

Therefore, the continuous spectra corresponding to these supersymmetric partner potentials (or Hamiltonians) coincide. $\qquad\square$

This example reflects general spectral properties of the supersymmetry. In our case, when $k > 0$ the eigenvalue equation (12.47) can be expressed in the supersymmetric formulation as

$$\mathcal{H}^{(1)}\Psi_k^{(1)} = \left[\mathcal{A}^\dagger\mathcal{A} + e_0\right]\Psi_k^{(1)} = e_k^{(1)}\Psi_k^{(1)}, \qquad (12.65)$$

which implies the eigenvalue equation

$$\mathcal{H}^{(2)}\left(\mathcal{A}\Psi_k^{(1)}\right) = \left[\mathcal{A}\mathcal{A}^\dagger\mathcal{A} + e_0\mathcal{A}\right]\Psi_k^{(1)} = e_k^{(1)}\left(\mathcal{A}\Psi_k^{(1)}\right). \qquad (12.66)$$

The same eigenvalue equation for $\mathcal{H}^{(2)}$,

$$\mathcal{H}^{(2)}\Psi_k^{(2)} = \left[\mathcal{A}\mathcal{A}^\dagger + e_0\right]\Psi_k^{(2)} = e_k^{(2)}\Psi_k^{(2)}, \qquad (12.67)$$

---

[9]A comprehensive account of this theory is provided in Ref. [198].

implies

$$\mathcal{H}^{(1)}\left(\mathcal{A}^\dagger \Psi_k^{(2)}\right) = [\mathcal{A}^\dagger \mathcal{A} \mathcal{A}^\dagger + e_0 \mathcal{A}^\dagger]\,\Psi_k^{(2)} = e_k^{(2)}\left(\mathcal{A}^\dagger \Psi_k^{(2)}\right). \qquad (12.68)$$

In the case of the discrete spectrum, these expressions establish relations between the eigenvalues and eigenfunctions of the supersymmetric partner Hamiltonians $\mathcal{H}^{(1)}$ and $\mathcal{H}^{(2)}$. Namely, we have for $n = 0, 1, 2, \ldots$ $e_n^{(2)} = e_{n+1}^{(1)}$, $e_0^{(1)} = 0$, while $\Psi_n^{(2)} = [e_{n+1}^{(1)}]^{-1/2} \mathcal{A} \Psi_n^{(1)}$ and $\Psi_{n+1}^{(1)} = [e_n^{(2)}]^{-1/2} \mathcal{A}^\dagger \Psi_n^{(2)}$. For the continuous spectrum, these supersymmetry expressions establish a relation between scattering parameters like the scattering amplitude, or transmission and reflection coefficients, which we consider as an example.

Example 12.2. Note that in order for scattering to take place in both of the partner potentials, it is necessary that the potentials are finite as $y \to -\infty$ or as $y \to \infty$ or both. This is fulfilled in our case since the pontentials are actually zero, see Example 12.1, at $y \to \pm\infty$. This also defines the asymptotic behavior of the superpotential,

$$W(y \to \pm\infty) = W_\pm = \pm\frac{\lambda}{2}, \qquad (12.69)$$

while the wave number defines the kinetic energy of the incident wave $e_k = k^2$. We write the asymptotic form of the solution $\Psi_k^+(y)$ in Eq. (12.52) as

$$\Psi_k^{(1,2)}(y \to \pm\infty) = e^{iky} + \mathcal{B}^{(1,2)}(k)e^{\pm iky}, \qquad (12.70)$$

and we find

$$\Psi_k^{(1)}(y \to \pm\infty) = \mathcal{A}^\dagger \Psi_k^{(2)}(y \to \pm\infty) = \left(-\frac{d}{dy} + W_\pm\right)\Psi_k^{(2)}(y \to \pm\infty). \qquad (12.71)$$

This yields

$$e^{iky} + \mathcal{B}^{(1)}(k)e^{-iky} = \left(-\frac{d}{dy} + W_-\right)\left(e^{iky} + \mathcal{B}^{(2)}(k)e^{-iky}\right)$$
$$= N\left[\left(-ik - \frac{\lambda}{2}\right)e^{iky} + \mathcal{B}^{(2)}(k)\left(ik - \frac{\lambda}{2}\right)e^{-iky}\right], \qquad (12.72)$$

$$\left(1 + \mathcal{B}^{(1)}(k)\right)e^{iky} = \left(-\frac{d}{dy} + W_+\right)\left(1 + \mathcal{B}^{(2)}(k)\right)e^{iky}$$
$$= N\left(-ik + \frac{\lambda}{2}\right)\left(1 + \mathcal{B}^{(2)}(k)\right)e^{iky}, \qquad (12.73)$$

where N is an overall normalization constant. Taking into account that $\mathcal{B}^{(1)} \equiv \mathcal{B}(k) = \frac{-\lambda}{\lambda+2ik}$, equating terms with the same exponents, and eliminating N, we find

$$\mathcal{B}^{(2)} = \frac{-\lambda}{\lambda - 2ik}. \tag{12.74}$$

This result can be verified straightforwardly by solving the eigenvalue equation for the $\mathcal{H}^{(2)}$ Hamiltonian. $\qquad\square$

Bibliography

[1] E. S. Swanson, *Science and Society: Understanding Scientific Methodology, Energy, Climate, and Sustainability.* Springer, Cham (2016), https://link.springer.com/book/10.1007%2F978-3-319-21987-5.

[2] E. W. Montroll and M. F. Shlesinger, On the wonderful world of random walks, in J. L. Lebowitz and E. W. Montroll (eds.), *Nonequilibrium Phenomena II: From Stochastics to Hydrodynamics.* Elsevier Science Publishers BV, Amsterdam, pp. 1–121 (1984).

[3] R. Metzler and J. Klafter, The random walk's guide to anomalous diffusion: A fractional dynamics approach, *Phys. Rep.* **339**, 1, pp. 1–77 (2000), http://dx.doi.org/10.1016/S0370-1573(00)00070-3.

[4] R. Metzler and J. Klafter, The restaurant at the end of the random walk: recent developments in the description of anomalous transport by fractional dynamics, *J. Phys. A: Math. Gen.* **37**, pp. R161–R208 (2004), http://dx.doi.org/10.1088/0305-4470/37/31/R01.

[5] E. W. Montroll and G. H. Weiss, Random walks on lattices. II, *J. Math. Phys.* **6**, 2, pp. 167–181 (1965), http://dx.doi.org/10.1063/1.1704269.

[6] K. Maex, M. R. Baklanov, D. Shamiryan, F. Iacopi, S. H. Brongersma and Z. S. Yanovitskaya, Low dielectric constant materials for microelectronics, *J. Appl. Phys.* **93**, 11, pp. 8793–8841 (2003), http://dx.doi.org/10.1063/1.1567460.

[7] E. F. Casassa and G. C. Berry, Angular distribution of intensity of Rayleigh scattering from comblike branched molecules, *J. Polymer Sci. A* **4**, 6, pp. 881–897 (1966), http://dx.doi.org/10.1002/pol.1966.160040605.

[8] J. F. Douglas, J. Roovers and K. F. Freed, Characterization of branching architecture through "universal" ratios of polymer solution properties, *Macromolecules* **23**, 18, pp. 4168–4180 (1990), http://dx.doi.org/10.1021/ma00220a022.

[9] H. Frauenrath, Dendronized polymers—building a new bridge from molecules to nanoscopic objects, *Prog. Polymer Sci.* **30**, 3–4, pp. 325–384 (2005), http://www.sciencedirect.com/science/article/pii/S0079670005000110.

[10] I. M. Sokolov, Models of anomalous diffusion in crowded environments, *Soft Matter* **8**, 35, pp. 9043–9052 (2012), http://dx.doi.org/10.1039/ C2SM25701G.

[11] T. A. L. Ziman, Excitations of dilute magnets near the percolation threshold, *J. Phys. C: Solid State Phys.* **12**, 13, pp. 2645–2658 (1979), http:// stacks.iop.org/0022-3719/12/i=13/a=027.

[12] S. R. White and M. Barma, Field-induced drift and trapping in percolation networks, *J. Phys. A: Math. Gen.* **17**, 15, pp. 2995–3008 (1984), http:// stacks.iop.org/0305-4470/17/i=15/a=017.

[13] G. H. Weiss and S. Havlin, Some properties of a random walk on a comb structure, *Physica A* **134**, 2, pp. 474–482 (1986), http://dx.doi.org/10. 1016/0378-4371(86)90060-9.

[14] Y. Gefen and I. Goldhirsch, Biased diffusion on random networks: mean first passage time and DC conductivity, *J. Phys. A: Math. Gen.* **18**, 16, pp. L1037–L1041 (1985), http://stacks.iop.org/0305-4470/18/i=16/ a=008.

[15] V. E. Arkhincheev and E. M. Baskin, Anomalous diffusion and drift in the comb model of percolation clusters, *Sov. Phys. JETP* **73**, 1, pp. 161–165 (1991), http://jetp.ac.ru/cgi-bin/e/index/e/73/1/p161?a=list.

[16] V. Méndez and A. Iomin, Comb-like models for transport along spiny dendrites, *Chaos, Solitons & Fractals* **53**, pp. 46–51 (2013), http://www. sciencedirect.com/science/article/pii/S0960077913000830.

[17] A. Iomin and V. Méndez, Reaction-subdiffusion front propagation in a comblike model of spiny dendrites, *Phys. Rev. E* **88**, 1, 012706 (2013), http://link.aps.org/doi/10.1103/PhysRevE.88.012706.

[18] R. E. Marsh, T. A. Riauka and S. A. McQuarrie, A review of basic principles of fractals and their application to pharmacokinetics, *Q. J. Nucl. Med. Mol. Imaging* **52**, 3, pp. 278–288 (2008), http://www.minervamedica.it/ index2.t?show=R39Y2008N03A0278.

[19] V. E. Arkhincheev, E. Kunnen and M. R. Baklanov, Active species in porous media: Random walk and capture in traps, *Microelectron. Eng.* **88**, 5, pp. 694–696 (2011), http://dx.doi.org/10.1016/j.mee.2010.08.028.

[20] H. E. Stanley and A. Coniglio, Flow in porous media: The "backbone" fractal at the percolation threshold, *Phys. Rev. B* **29**, 1, pp. 522–524 (1984), http://link.aps.org/doi/10.1103/PhysRevB.29.522.

[21] A. Tarasenko and L. Jastrabík, A one-dimensional lattice-gas model for simulating diffusion in channel pores with side pockets: The analytical approach and kinetic Monte Carlo technique, *Microporous and Mesoporous Materials* **152**, pp. 134–140 (2012), http://www.sciencedirect. com/science/article/pii/S1387181111005725.

[22] S. Hannunen, Vegetation architecture and redistribution of insects moving on the plant surface, *Ecol. Model.* **155**, 2–3, pp. 149–157 (2002), http:// www.sciencedirect.com/science/article/pii/S0304380002001254.

[23] N. Korabel and E. Barkai, Paradoxes of subdiffusive infiltration in disordered systems, *Phys. Rev. Lett.* **104**, 17, 170603 (2010), http://link.aps. org/doi/10.1103/PhysRevLett.104.170603.

[24] E. Baskin and A. Iomin, Superdiffusion on a comb structure, *Phys. Rev. Lett.* **93**, 12, 120603 (2004), http://dx.doi.org/10.1103/PhysRevLett.93.120603.

[25] A. Compte, Continuous time random walks on moving fluids, *Phys. Rev. E* **55**, 6, pp. 6821–6831 (1997), http://link.aps.org/abstract/PRE/v55/p6821.

[26] A. Iomin and E. Baskin, Negative superdiffusion due to inhomogeneous convection, *Phys. Rev. E* **71**, 6, 061101 (2005), http://dx.doi.org/10.1103/PhysRevE.71.061101.

[27] R. Metzler, E. Barkai and J. Klafter, Anomalous diffusion and relaxation close to thermal equilibrium: A fractional Fokker-Planck equation approach, *Phys. Rev. Lett.* **82**, 18, pp. 3563–3567 (1999), http://dx.doi.org/10.1103/PhysRevLett.82.3563.

[28] R. Metzler and A. Compte, Generalized diffusion-advection schemes and dispersive sedimentation: A fractional approach, *J. Phys. Chem. B* **104**, 16, pp. 3858–3865 (2000), http://dx.doi.org/10.1021/jp993698f.

[29] C. Van den Broeck, Waiting times for random walks on regular and fractal lattices, *Phys. Rev. Lett.* **62**, 13, pp. 1421–1424 (1989), http://link.aps.org/abstract/PRL/v62/p1421.

[30] D. Campos and V. Méndez, Reaction-diffusion wave fronts on comblike structures, *Phys. Rev. E* **71**, 5, 051104 (2005), http://dx.doi.org/10.1103/PhysRevE.71.051104.

[31] D. Campos, J. Fort and V. Méndez, Transport on fractal river networks: Application to migration fronts, *Theor. Pop. Biol.* **69**, 1, pp. 88–93 (2006), http://dx.doi.org/10.1016/j.tpb.2005.09.001.

[32] E. Agliari, A. Blumen and D. Cassi, Slow encounters of particle pairs in branched structures, *Phys. Rev. E* **89**, 5, 052147 (2014), http://link.aps.org/doi/10.1103/PhysRevE.89.052147.

[33] A. Rebenshtok and E. Barkai, Occupation times on a comb with ramified teeth, *Phys. Rev. E* **88**, 5, 052126 (2013), http://link.aps.org/doi/10.1103/PhysRevE.88.052126.

[34] G. Forte, R. Burioni, F. Cecconi and A. Vulpiani, Anomalous diffusion and response in branched systems: a simple analysis, *J. Phys.: Condens. Matter* **25**, 46, 465106 (2013), http://stacks.iop.org/0953-8984/25/i=46/a=465106.

[35] G. H. Weiss, *Aspects and Applications of the Random Walk*. North-Holland, Amsterdam (1994).

[36] D. ben Avraham and S. Havlin, *Diffusion and Reactions in Fractals and Disordered Systems*. Cambridge University Press, Cambridge, UK (2000).

[37] J. Klafter and I. M. Sokolov, *First Steps in Random Walks: From Tools to Applications*. Oxford University Press, New York (2011).

[38] D. R. Cox, *Renewal Theory*. Methuen, London (1967).

[39] K. B. Oldham and J. Spanier, *The Fractional Calculus: Theory and Applications of Differentiation and Integration to Arbitrary Order*. Academic Press, New York (1974).

[40] K. Miller and B. Ross, *An Introduction to the Fractional Calculus and Fractional Differential Equations*. John Wiley & Sons, New York (1993).

[41] I. Podlubny, *Fractional Differential Equations*. Academic Press, San Diego (1999).

[42] S. G. Samko, A. A. Kilbas and O. I. Marichev, *Fractional Integrals and Derivatives: Theory and Applications*. Gordon and Breach, London (1993).

[43] H. Bateman and A. Erdélyi, *Higher Transcendental Functions [Volumes I-III]*. McGraw-Hill, New York (1953–1955), http://authors.library.caltech.edu/43491/.

[44] A. M. Mathai, R. K. Saxena and H. J. Haubold, *The H-function: Theory and Applications*. Springer, New York (2010), http://dx.doi.org/10.1007/978-1-4419-0916-9.

[45] A. M. Mathai and H. J. Haubold, *Special Functions for Applied Scientists*. Springer, New York (2008), http://link.springer.com/book/10.1007/978-0-387-75894-7.

[46] M. Abramovitz and I. A. Stegun, *Handbook of Mathematical Functions with Formulas, Graphs, and Mathematical Tables*. Dover Publications, New York (1972).

[47] G. M. Fikhtengoltz, *Course of Differential and Integral Calculus*, Vol. 2. Nauka, Moscow (1969).

[48] F. Oberhettinger, *Tables of Mellin Transforms*. Springer, Berlin (1974).

[49] V. Méndez, D. Campos and F. Bartumeus, *Stochastic Foundations in Movement Ecology: Anomalous Diffusion, Front Propagation and Random Searches*. Springer, Berlin (2014), http://link.springer.com/book/10.1007/978-3-642-39010-4/page/1.

[50] W. Feller, *An Introduction to Probability Theory and its Applications*, Vol. 1. Wiley, New York (1971).

[51] W. Feller, *An Introduction to Probability Theory and its Applications*, Vol. 2. Wiley, New York (1971).

[52] N. G. Van Kampen, *Stochastic Processes in Physics and Chemistry*, 3rd edn. North Holland, Amsterdam (2007), http://www.sciencedirect.com/science/book/9780444529657.

[53] S. M. Rytov, Y. A. Kravtsov and V. I. Tatarskii, *Principles of Statistical Radiophysics 1: Elements of Random Process Theory*. Springer-Verlag, Berlin (1987), http://www.springer.com/us/book/9783642692031.

[54] S. Karlin and H. M. Taylor, *A First Course in Stochastic Processes*, 2nd edn. Academic Press, Boston (1975), http://www.sciencedirect.com/science/article/pii/B9780080570419500015.

[55] C. W. Gardiner, *Handbook of Stochastic Methods for Physics, Chemistry and the Natural Sciences*, 2nd edn. Springer-Verlag, Berlin (1990).

[56] A. J. Lichtenberg and M. A. Lieberman, *Regular and Chaotic Dynamics*. Springer, New York (1983).

[57] E. W. Montroll and B. J. West, Chapter 2 – On an Enriched Collection of Stochastic Processes, in E. W. Montroll and J. L. Lebowitz (eds.), *Fluctuation Phenomena*. Elsevier, Amsterdam, pp. 61–175 (1979), http://www.sciencedirect.com/science/article/pii/B9780444852489500054.

[58] M. Kac, *Some Stochastic Problems in Physics and Mathematics, Colloquium Lectures in Pure and Applied Science*, Vol. 2. Magnolia

Petroleum Company, Dallas, TX (1956), https://books.google.com/books?id=llarGAAACAAJ.

[59] M. C. Wang and G. E. Uhlenbeck, On the theory of the Brownian motion II, *Rev. Mod. Phys.* **17**, 2–3, pp. 323–342 (1945), http://dx.doi.org/10.1103/RevModPhys.17.323.

[60] P. Lévy, *Théorie de l'addition des variables aléatoires*, 2nd edn. Gauthier-Villars, Paris (1954).

[61] V. Uchaikin and R. Sibatov, *Fractional Kinetics in Solids*. World Scientific, Singapore (2013), http://www.worldscientific.com/worldscibooks/10.1142/8185.

[62] H. Bateman and A. Erdélyi, *Tables of Integral Transforms [Volumes I & II]*. McGraw-Hill, New York (1954), http://resolver.caltech.edu/CaltechAUTHORS:20140123-101456353.

[63] L. Arnold, *Stochastic Differential Equations: Theory and Applications*. John Wiley & Sons, New York (1974).

[64] W. Horsthemke and R. Lefever, *Noise-Induced Transitions. Theory and Applications in Physics, Chemistry, and Biology*. Springer-Verlag, Berlin (1984), http://dx.doi.org/10.1007/3-540-36852-3.

[65] R. F. Pawula, Approximation of the linear Boltzmann equation by the Fokker-Planck equation, *Phys. Rev.* **162**, 1, pp. 186–188 (1967), https://link.aps.org/doi/10.1103/PhysRev.162.186.

[66] P. Langevin, Sur la théorie du mouvement brownien, *C. R. Acad. Sci. Paris* **146**, pp. 530–533 (1908).

[67] D. S. Lemons and A. Gythiel, Paul Langevin's 1908 paper "On the theory of Brownian motion" ["Sur la théorie du mouvement brownien," C. R. Acad. Sci. (Paris) 146, 530–533 (1908)], *Am. J. Phys.* **65**, 11, pp. 1079–1081 (1997), https://doi.org/10.1119/1.18725.

[68] K. Ito, Stochastic differential equations in a differentiable manifold, *Nagoya Math. J.* **1**, pp. 35–47 (1950), http://projecteuclid.org/euclid.nmj/1118764702.

[69] R. L. Stratonovich, A new representation for stochastic integrals and equations, *SIAM J. Control* **4**, 2, pp. 362–371 (1966), http://dx.doi.org/10.1137/0304028.

[70] Z. Schuss, *Theory and Applications of Stochastic Processes*. Springer, New York (2010), http://dx.doi.org/10.1007/978-1-4419-1605-1.

[71] V. Méndez, S. I. Denisov, D. Campos and W. Horsthemke, Role of the interpretation of stochastic calculus in systems with cross-correlated Gaussian white noises, *Phys. Rev. E* **90**, 1, 012116 (2014), http://link.aps.org/doi/10.1103/PhysRevE.90.012116.

[72] P. Hänggi, Stochastic processes I: Asymptotic behaviour and symmetries, *Helv. Phys. Acta* **51**, pp. 183–201 (1978), http://www.physik.uni-augsburg.de/theo1/hanggi/Papers/7.pdf.

[73] Y. L. Klimontovich, Ito, Stratonovich and kinetic forms of stochastic equations, *Physica A* **163**, 2, pp. 515–532 (1990), http://dx.doi.org/10.1016/0378-4371(90)90142-F.

[74] Y. L. Klimontovich, *Statistical Theory of Open Systems*. Kluwer Academic, Dordrecht (1995).

[75] E. Wong and M. Zakai, On the convergence of ordinary integrals to stochastic integrals, *Ann. Math. Stat.* **36**, 5, pp. 1560–1564 (1965), http://www.jstor.org/stable/2238444.

[76] S. I. Denisov, W. Horsthemke and P. Hänggi, Generalized Fokker-Planck equation: Derivation and exact solutions, *Eur. Phys. J. B* **68**, 4, p. 567575 (2009), http://dx.doi.org/10.1140/epjb/e2009-00126-3.

[77] S. Redner, *A Guide to First-Passage Processes*. Cambridge University Press, Cambridge, MA (2001), https://books.google.com/books?id=xtsqMh3VC98C.

[78] E. S. Andersen, On the fluctuations of sums of random variables, *Math. Scand.* **1**, 2, pp. 263–285 (1953), http://dx.doi.org/10.7146/math.scand.a-10385.

[79] G. Pólya, Über eine Aufgabe der Wahrscheinlichkeitsrechnung betreffend die Irrfahrt im Straßennetz, *Math. Ann.* **84**, 1, pp. 149–160 (1921), https://doi.org/10.1007/BF01458701.

[80] B. D. Hughes, *Random Walks and Random Environments: Volume 1: Random Walks*. Clarendon Press, Oxford (1995).

[81] B. Kahng and S. Redner, Scaling of the first-passage time and the survival probability on exact and quasi-exact self-similar structures, *J. Phys. A: Math. Gen.* **22**, 7, pp. 887–902 (1989), http://stacks.iop.org/0305-4470/22/i=7/a=019.

[82] L. Dagdug, A. M. Berezhkovskii, Y. A. Makhnovskii and V. Y. Zitserman, Transient diffusion in a tube with dead ends, *J. Chem. Phys.* **127**, 22, 224712 (2007), http://link.aip.org/link/?JCP/127/224712/1.

[83] V. Méndez, S. Fedotov and W. Horsthemke, *Reaction-Transport Systems: Mesoscopic Foundations, Fronts, and Spatial Instabilities*. Springer, Heidelberg (2010), http://dx.doi.org/10.1007/978-3-642-11443-4.

[84] S. I. Denisov and W. Horsthemke, Anomalous diffusion of particles driven by correlated noise, *Phys. Rev. E* **62**, 6, pp. 7729–7734 (2000), http://dx.doi.org/10.1103/PhysRevE.62.7729.

[85] V. Méndez, A. Iomin, D. Campos and W. Horsthemke, Mesoscopic description of random walks on combs, *Phys. Rev. E* **92**, 6, 062112 (2015), http://link.aps.org/doi/10.1103/PhysRevE.92.062112.

[86] I. Goldhirsch and Y. Gefen, Analytic method for calculating properties of random walks on networks, *Phys. Rev. A* **33**, 4, pp. 2583–2594 (1986), http://link.aps.org/doi/10.1103/PhysRevA.33.2583.

[87] I. Goldhirsch and Y. Gefen, Biased random walk on networks, *Phys. Rev. A* **35**, 3, pp. 1317–1327 (1987), http://link.aps.org/doi/10.1103/PhysRevA.35.1317.

[88] V. Balakrishnan and C. Van den Broeck, Transport properties on a random comb, *Physica A* **217**, 1–2, pp. 1–21 (1995), http://www.sciencedirect.com/science/article/pii/037843719500083J.

[89] S. Havlin, J. E. Kiefer and G. H. Weiss, Anomalous diffusion on a random comblike structure, *Phys. Rev. A* **36**, 3, pp. 1403–1408 (1987), http://link.aps.org/doi/10.1103/PhysRevA.36.1403.

[90] V. I. Klyatskin, *Stochastic Equations through the Eye of the Physicist: Basic Concepts, Exact Results and Asymptotic Approximations*. Elsevier Science, Amsterdam (2005), https://books.google.com/books?id=nntqG3KCSVkC.

[91] K. Furutsu, On the theory of radio wave propagation over inhomogeneous earth, *J. Res. Nat. Bur. Stand* **67**, pp. 39–62 (1963).

[92] E. A. Novikov, Functionals and the random-force method in turbulence, *Zh. Eksp. Teor. Fiz.* **47**, p. 1919 (1964), [Sov. Phys. JETP **20**, 1990 (1965)].

[93] G. Bel and I. Nemenman, Ergodic and non-ergodic anomalous diffusion in coupled stochastic processes, *New J. Phys.* **11**, 8, 083009 (2009), http://stacks.iop.org/1367-2630/11/i=8/a=083009.

[94] H. V. Ribeiro, A. A. Tateishi, L. G. A. Alves, R. S. Zola and E. K. Lenzi, Investigating the interplay between mechanisms of anomalous diffusion via fractional Brownian walks on a comb-like structure, *New J. Phys.* **16**, 9, 093050 (2014), http://stacks.iop.org/1367-2630/16/i=9/a=093050.

[95] V. Méndez, A. Iomin, W. Horsthemke and D. Campos, Langevin dynamics for ramified structures, *J. Stat. Mech.: Theor. Exp.* **2017**, 6, 063205 (2017), http://stacks.iop.org/1742-5468/2017/i=6/a=063205.

[96] V. Volterra, *Theory of Functionals*. Blackie and Son, London (1930).

[97] V. I. Klyatskin, *Stohasticheskie uravneniya i volny v sluchaino-neodnorodnyh sredah (Stochastic equations and waves in randomly inhomogeneous media)*. Nauka, Moscow (1980).

[98] J. Zinn-Justin, *Quantum Field Theory and Critical Phenomena*. Clarendon Press, Oxford (1990).

[99] V. I. Tatarskii, *Rasprostranenie Voln v Turbulentnoi Atmosfere (Wave Propagation in Turbulent Atmosphere)*. Nauka, Moscow (1967).

[100] P. Hänggi and P. Jung, Colored noise in dynamical systems, in I. Prigogine and S. A. Rice (eds.), *Advances in Chemical Physics*, Vol. 89. John Wiley & Sons, Inc., New York, pp. 239–326 (1995), http://dx.doi.org/10.1002/9780470141489.ch4.

[101] A. V. Chechkin, V. Y. Gonchar, J. Klafter, R. Metzler and L. V. Tanatarov, Lévy flights in a steep potential well, *J. Stat. Phys.* **115**, 5, pp. 1505–1535 (2004), http://dx.doi.org/10.1023/B:JOSS.0000028067.63365.04.

[102] B. Berkowitz, J. Klafter, R. Metzler and H. Scher, Physical pictures of transport in heterogeneous media: Advection-dispersion, random-walk, and fractional derivative formulations, *Water Resour. Res.* **38**, 10, 1191 (2002), http://dx.doi.org/10.1029/2001WR001030.

[103] R. Hilfer, Transport and relaxation phenomena in porous media, in I. Prigogine and S. A. Rice (eds.), *Advances in Chemical Physics*. John Wiley & Sons, Inc., New York, pp. 299–424 (2007), http://dx.doi.org/10.1002/9780470141519.ch3.

[104] J. S. Andrade, Jr., D. A. Street, Y. Shibusa, S. Havlin and H. E. Stanley, Diffusion and reaction in percolating pore networks, *Phys. Rev. E* **55**, 1, pp. 772–777 (1997), http://link.aps.org/abstract/PRE/v55/p772.

[105] B. J. West, P. Grigolini, R. Metzler and T. F. Nonnenmacher, Fractional diffusion and Lévy stable processes, *Phys. Rev. E* **55**, 1, pp. 99–106 (1997), http://link.aps.org/doi/10.1103/PhysRevE.55.99.

[106] T. F. Nonnenmacher and R. Metzler, Applications of fractional calculus techniques to problems in biophysics, in R. Hilfer (ed.), *Applications of Fractional Calculus in Physics*. World Scientific, Singapore, pp. 377–427 (2000), http://dx.doi.org/10.1142/9789812817747_0008.

[107] J. Klafter, A. Blumen and M. F. Shlesinger, Stochastic pathway to anomalous diffusion, *Phys. Rev. A* **35**, 7, pp. 3081–3085 (1987), http://dx.doi.org/10.1103/PhysRevA.35.3081.

[108] I. M. Sokolov and R. Metzler, Towards deterministic equations for Lévy walks: The fractional material derivative, *Phys. Rev. E* **67**, 1, 010101(R) (2003), http://dx.doi.org/10.1103/PhysRevE.67.010101.

[109] A. Chechkin, V. Gonchar, J. Klafter, R. Metzler and L. Tanatarov, Stationary states of non-linear oscillators driven by Lévy noise, *Chem. Phys.* **284**, 1–2, pp. 233–251 (2002), http://www.sciencedirect.com/science/article/pii/S0301010402005517.

[110] F. W. J. Olver, *Introduction to Asymptotics and Special Functions*. Acaddemic Press, New York (1974).

[111] A. Erdélyi, *Asymptotic Expansions*. Dover Publications, New York (1956), https://books.google.com/books?id=aedk-OHdmNYC.

[112] B. O'Shaughnessy and I. Procaccia, Analytical solutions for diffusion on fractal objects, *Phys. Rev. Lett.* **54**, 5, pp. 455–458 (1985), http://link.aps.org/abstract/PRL/v54/p455.

[113] M. A. A. da Silva, G. M. Viswanathan and J. C. Cressoni, Ultraslow diffusion in an exactly solvable non-Markovian random walk, *Phys. Rev. E* **89**, 5, 052110 (2014), http://link.aps.org/doi/10.1103/PhysRevE.89.052110.

[114] D. Boyer and C. Solis-Salas, Random walks with preferential relocations to places visited in the past and their application to biology, *Phys. Rev. Lett.* **112**, 24, 240601 (2014), http://link.aps.org/doi/10.1103/PhysRevLett.112.240601.

[115] A. V. Chechkin, J. Klafter and I. M. Sokolov, Fractional Fokker-Planck equation for ultraslow kinetics, *Europhys. Lett.* **63**, 3, pp. 326–332 (2003), http://dx.doi.org/10.1209/epl/i2003-00539-0.

[116] S. I. Denisov and H. Kantz, Continuous-time random walk theory of superslow diffusion, *Europhys. Lett.* **92**, 3, 30001 (2010), http://stacks.iop.org/0295-5075/92/i=3/a=30001.

[117] A. Godec, A. V. Chechkin, E. Barkai, H. Kantz and R. Metzler, Localisation and universal fluctuations in ultraslow diffusion processes, *J. Phys. A: Math. Theor.* **47**, 49, 492002 (2014), http://stacks.iop.org/1751-8121/47/i=49/a=492002.

[118] A. S. Bodrova, A. V. Chechkin, A. G. Cherstvy and R. Metzler, Ultraslow scaled Brownian motion, *New J. Phys.* **17**, 6, 063038 (2015), http://stacks.iop.org/1367-2630/17/i=6/a=063038.

[119] J.-P. Bouchaud and A. Georges, Anomalous diffusion in disordered media: Statistical mechanisms, models and physical applications, *Phys. Rep.* **195**, 4-5, pp. 127–293 (1990), http://dx.doi.org/10.1016/0370-1573(90)90099-N.

[120] S. Alexander and R. Orbach, Density of states on fractals: "fractons", *J. Physique Lett.* **43**, 17, pp. 625–631 (1982), http://dx.doi.org/10.1051/jphyslet:019820043017062500.

[121] A. Iomin and V. Méndez, Does ultra-slow diffusion survive in a three dimensional cylindrical comb? *Chaos, Solitons & Fractals* **82**, pp. 142–147 (2016), http://www.sciencedirect.com/science/article/pii/S0960077915003823.

[122] A. R. Dzhanoev and I. M. Sokolov, The effect of the junction model on the anomalous diffusion in the 3D comb structure, *Chaos, Solitons & Fractals* **106**, pp. 330–336 (2018), https://www.sciencedirect.com/science/article/pii/S0960077917305039.

[123] O. A. Dvoretskaya, P. S. Kondratenko and L. V. Matveev, Anomalous diffusion in generalized Dykhne model, *JETP* **110**, 1, pp. 58–66 (2010), http://dx.doi.org/10.1134/S1063776110010085.

[124] E. Jahnke, F. Emde and F. Lösch, *Tables of Higher Functions.* McGraw-Hill, New York (1960).

[125] I. S. Gradshteyn and I. M. Ryzhik, *Table of Integrals, Series, and Products.* Elsevier Academic Press, Amsterdam (2007).

[126] G. Zaslavsky, Fractional kinetic equation for Hamiltonian chaos, *Physica D* **76**, 1-3, pp. 110–122 (1994), http://dx.doi.org/10.1016/0167-2789(94)90254-2.

[127] A. Godec and R. Metzler, Linear response, fluctuation-dissipation, and finite-system-size effects in superdiffusion, *Phys. Rev. E* **88**, 1, 012116 (2013), http://link.aps.org/doi/10.1103/PhysRevE.88.012116.

[128] A. Iomin, V. Zaburdaev and T. Pfohl, Reaction front propagation of actin polymerization in a comb-reaction system, *Chaos, Solitons & Fractals* **92**, pp. 115–122 (2016), http://www.sciencedirect.com/science/article/pii/S096007791630265X.

[129] S. B. Yuste, E. Abad and A. Baumgaertner, Anomalous diffusion and dynamics of fluorescence recovery after photobleaching in the random-comb model, *Phys. Rev. E* **94**, 012118 (2016), http://link.aps.org/doi/10.1103/PhysRevE.94.012118.

[130] J. Rose, S.-X. Jin and A. Craig, Heterosynaptic molecular dynamics: locally induced propagating synaptic accumulation of CaM kinase II, *Neuron* **61**, 3, pp. 351–358 (2009), http://www.sciencedirect.com/science/article/pii/S0896627309000361.

[131] B. A. Earnshaw and P. C. Bressloff, A diffusion-activation model of CaMKII translocation waves in dendrites, *J. Comput. Neurosci.* **28**, pp. 77–89 (2010), http://dx.doi.org/10.1007/s10827-009-0188-9.

[132] R. Yuste, *Dendritic Spines.* MIT Press (2010), http://books.google.es/books?id=reRPgWOImqIC.

[133] E. Korkotian and M. Segal, Spatially confined diffusion of calcium in dendrites of hippocampal neurons revealed by flash photolysis of caged calcium, *Cell Calcium* **40**, pp. 441–449 (2006), http://www.sciencedirect.com/science/article/pii/S0143416006001734.

[134] M. Segal, Dendritic spines and long-term plasticity, *Nature Rev. Neurosci.* **6**, pp. 277–284 (2005), http://dx.doi.org/10.1038/nrn1649.

[135] S. Coombes and P. C. Bressloff, Saltatory waves in the spike-diffuse-spike model of active dendritic spines, *Phys. Rev. Lett.* **91**, 028102 (2003), http://link.aps.org/doi/10.1103/PhysRevLett.91.028102.

[136] R. A. Fisher, The wave of advance of advantageous genes, *Ann. Eugenics* **7**, pp. 353–369 (1937), http://dx.doi.org/10.1111/j.1469-1809.1937.tb02153.x.

[137] A. Kolmogorov, I. Petrovskii and N. Piskunov, Étude de l'équation de la diffusion avec croissance de la quantité de matière et son application à un problème biologique (Study of a diffusion equation that is related to the growth of a quantity of matter and its application to a biological problem), *Mosc. Univ. Bull. Math. Ser. A* **1**, pp. 1–25 (1937).

[138] E. A. Nimchinsky, B. L. Sabatini and K. Svoboda, Structure and function of dendritic spines, *Annu. Rev. Physiol.* **64**, pp. 313–353 (2002), http://dx.doi.org/10.1146/annurev.physiol.64.081501.160008.

[139] A. M. Berezhkovskii, A. V. Barzykin and V. Y. Zitserman, Escape from cavity through narrow tunnel, *J. of Chem. Phys.* **130**, 24, 245104 (2009), http://dx.doi.org/10.1063/1.3160546.

[140] M. J. Byrne, M. N. Waxham and Y. Kubota, The impacts of geometry and binding on CaMKII diffusion and retention in dendritic spines, *J. Comput. Neurosci.* **31**, pp. 1–12 (2011), http://dx.doi.org/10.1007/s10827-010-0293-9.

[141] F. Santamaria, S. Wils, E. De Schutter and G. J. Augustine, Anomalous diffusion in Purkinje cell dendrites caused by spines, *Neuron* **52**, 4, pp. 635–648 (2006), http://dx.doi.org/10.1016/j.neuron.2006.10.025.

[142] F. Santamaria, S. Wils, E. De Schutter and G. J. Augustine, The diffusional properties of dendrites depend on the density of dendritic spines, *Eur. J. Neurosci.* **34**, 4, pp. 561–568 (2011), http://dx.doi.org/10.1111/j.1460-9568.2011.07785.x.

[143] S. Zeng and W. R. Holmes, The effect of noise on CaMKII activation in a dendritic spine during LTP induction, *J. Neurophysiol.* **103**, 4, pp. 1798–1808 (2010), http://jn.physiology.org/content/103/4/1798.

[144] D. Froemberg, H. Schmidt-Martens, I. M. Sokolov and F. Sagués, Front propagation in A + B → 2 A reaction under subdiffusion, *Phys. Rev. E* **78**, 1, 011128 (2008), http://link.aps.org/abstract/PRE/v78/e011128.

[145] P. C. Bressloff, Propagation of CaMKII translocation waves in heterogeneous spiny dendrites, *J. Math. Biol.* **66**, 7, pp. 1499–1525 (2013), http://dx.doi.org/10.1007/s00285-012-0542-9.

[146] A. Iomin, Toy model of fractional transport of cancer cells due to self-entrapping, *Phys. Rev. E* **73**, 6, 061918 (2006), http://link.aps.org/abstract/PRE/v73/e061918.

[147] M. Freidlin, *Markov Processes and Differential Equations*. Birkhäuser, Basel (1996).

[148] S. Fedotov and V. Méndez, Non-Markovian model for transport and reactions of particles in spiny dendrites, *Phys. Rev. Lett.* **101**, 21, 218102 (2008), http://link.aps.org/abstract/PRL/v101/e218102.

[149] S. Fedotov, Non-Markovian random walks and nonlinear reactions: Subdiffusion and propagating fronts, *Phys. Rev. E* **81**, 011117 (2010), http://link.aps.org/doi/10.1103/PhysRevE.81.011117.

[150] S. Deshpande and T. Pfohl, Hierarchical self-assembly of actin in micro-confinements using microfluidics, *Biomicrofluidics* **6**, 3, 034120 (2012), http://scitation.aip.org/content/aip/journal/bmf/6/3/10. 1063/1.4752245.

[151] S. Deshpande and T. Pfohl, Real-time dynamics of emerging actin networks in cell-mimicking compartments, *PLoS ONE* **10**, 3, e0116521 (2015), http://dx.doi.org/10.1371/journal.pone.0116521.

[152] S. Köster, J. Kierfeld and T. Pfohl, Characterization of single semiflexible filaments under geometric constraints, *Eur. Phys. J. E* **25**, 4, pp. 439–449 (2008), http://dx.doi.org/10.1140/epje/i2007-10312-3.

[153] D. Froemberg, H. H. Schmidt-Martens, I. M. Sokolov and F. Sagués, Asymptotic front behavior in an $A + B \to 2\,A$ reaction under subdiffusion, *Phys. Rev. E* **83**, 3, 031101 (2011), http://link.aps.org/doi/10. 1103/PhysRevE.83.031101.

[154] A. Iomin and I. M. Sokolov, Application of hyperbolic scaling for calculation of reaction-subdiffusion front propagation, *Phys. Rev. E* **86**, 2, 022101 (2012), http://link.aps.org/doi/10.1103/PhysRevE.86.022101.

[155] H. S. Fogler, *Elements of Chemical Reaction Engineering*. Prentice Hall, Upper Saddle River, NJ (2006).

[156] H. J. Haubold, A. M. Mathai and R. K. Saxena, Mittag-Leffler functions and their applications, *J. Appl. Math.* **2011**, e298628 (2011), http://www. hindawi.com/journals/jam/2011/298628/abs/.

[157] T. Sandev, A. Iomin and H. Kantz, Fractional diffusion on a fractal grid comb, *Phys. Rev. E* **91**, 3, 032108 (2015), http://link.aps.org/doi/10. 1103/PhysRevE.91.032108.

[158] B. Mandelbrot, *The Fractal Geometry of Nature*. W.H. Freeman and Company, New York (1983).

[159] A. N. Kolmogoroff, Wienersche Spiralen und einige andere interessante Kurven im Hilbertschen Raum. *C. R. (Dokl.) Acad. Sci. URSS, n. Ser.* **26**, pp. 115–118 (1940).

[160] B. B. Mandelbrot and J. W. V. Ness, Fractional Brownian motions, fractional noises and applications, *SIAM Review* **10**, 4, pp. 422–437 (1968), http://dx.doi.org/10.1137/1010093, http://dx.doi.org/10. 1137/1010093.

[161] K. J. Falconer, *Fractal Geometry*. John Wiley & Sons Ltd, New York (1990).

[162] A. Lasota and M. Mackey, *Fractals and Noise, Stochastic Aspects of Dynamics*. Springer-Verlag, New York (1994).

[163] K. T. Alligood, T. D. Sauer and J. A. Yorke, *Chaos. An Introduction to Dynamical Systems*. Springer, New York (2000).

[164] T. Sandev, A. Iomin, H. Kantz, R. Metzler and A. Chechkin, Comb model with slow and ultraslow diffusion, *Math. Model. Nat. Phenom.* **11**, 3, pp. 18–33 (2016), http://dx.doi.org/10.1051/mmnp/201611302.

[165]   T. Sandev, A. Iomin and V. Méndez, Lévy processes on a generalized fractal comb, *J. Phys. A: Math. Theor.* **49**, 355001 (2016), http://stacks.iop. org/1751-8121/49/i=35/a=355001.

[166]   A. Iomin, Subdiffusion on a fractal comb, *Phys. Rev. E* **83**, 5, 052106 (2011), http://dx.doi.org/10.1103/PhysRevE.83.052106.

[167]   M. F. Shlesinger, Asymptotic solutions of continuous-time random walks, *J. Stat. Phys.* **10**, 5, pp. 421–434 (1974), http://dx.doi.org/10.1007/ BF01008803.

[168]   R. Schilling, R. Song and Z. Vondracek, *Bernstein Functions*. De Gruyter, Berlin (2010).

[169]   C.-S. Deng and R. L. Schilling, Complete Bernstein functions and subordinators with nested ranges. A note on a paper by P. Marchal, *Electron. Commun. Probab.* **21**, 78 (2016), http://dx.doi.org/10.1214/16-ECP31.

[170]   C. Berg and G. Forst, *Potential Theory on Locally Compact Abelian Groups*. Springer, Berlin (1975).

[171]   R. P. Feynman and A. R. Hibbs, *Quantum Mechanics and Path Integrals*. McGraw-Hill, New York (1965).

[172]   M. Kac, *Probability and Related Topics in Physical Sciences*. Interscience Publishers Inc., New York (1959).

[173]   N. Laskin, Fractals and quantum mechanics, *Chaos* **10**, 4, pp. 780–790 (2000a), http://dx.doi.org/10.1063/1.1050284.

[174]   N. Laskin, Fractional quantum mechanics, *Phys. Rev. E* **62**, 3, pp. 3135– 3145 (2000b), https://link.aps.org/doi/10.1103/PhysRevE.62.3135.

[175]   N. Laskin, Fractional quantum mechanics and Lévy path integrals, *Phys. Lett. A* **268**, 4, pp. 298–305 (2000c), doi:10.1016/S0375-9601(00)00201-2, http://www.sciencedirect.com/science/article/pii/S037596010000 2012.

[176]   N. Laskin, Fractional Schrödinger equation, *Phys. Rev. E* **66**, 5, p. 056108 (2002), https://link.aps.org/doi/10.1103/PhysRevE.66.056108.

[177]   B. J. West, Quantum Lévy propagators, *J. Phys. Chem. B* **104**, 16, pp. 3830–3832 (2000), http://dx.doi.org/10.1021/jp993323u.

[178]   M. Naber, Time fractional Schrödinger equation, *J. Math. Phys.* **45**, 8, pp. 3339–3352 (2004), http://dx.doi.org/10.1063/1.1769611.

[179]   M. H. Stone, On one-parameter unitary groups in Hilbert space, *Ann. Math.* **33**, 3, pp. 643–648 (1932), http://www.jstor.org/stable/1968538.

[180]   B. Gaveau and L. S. Schulman, Explicit time-dependent Schrödinger propagators, *J. Phys. A: Math. Gen.* **19**, 10, pp. 1833–1846 (1986), http:// stacks.iop.org/0305-4470/19/i=10/a=024.

[181]   L. Schulman, *Techniques and Applications of Path Integration*. Wiley, New York (1981).

[182]   J. V. Armitage, Chapter 10 – The Riemann hypothesis and the Hamiltonian of a quantum mechanical system, in M. M. Dodson and J. A. G. Vickers (eds.), *Number Theory and Dynamical Systems*, London Mathematical Society Lecture Note Series, Vol. 134. Cambridge University Press, Cambridge, UK, pp. 153–172 (1989), https://dx.doi.org/10. 1017/CBO9780511661983.011.

[183]   M. V. Berry and J. P. Keating, $h = xp$ and the Riemann zeros, in I. V. Lerner, J. P. Keating and D. E. Khmelnitskii (eds.), *Supersymmetry*

and Trace Formulae: Chaos and Disorder, NATO Advanced Studies Institute, Series B Physics, Vol. 370. Springer, New York, pp. 355–367 (1999), http://link.springer.com/chapter/10.1007/978-1-4615-4875-1_19.

[184] G. Berman and M. Vishik, Long time evolution of quantum averages near stationary points, *Phys. Lett. A* **319**, 3–4, pp. 352–359 (2003), http://www.sciencedirect.com/science/article/pii/S0375960103015688.

[185] A. Iomin, Exponential spreading and singular behavior of quantum dynamics near hyperbolic points, *Phys. Rev. E* **87**, 5, 054901 (2013), https://link.aps.org/doi/10.1103/PhysRevE.87.054901.

[186] V. A. Fock, *Foundations of Quantum Mechanics (in Russian)*. Nauka, Moscow (1976).

[187] K. L. Sebastian, Path integral representation for fractional Brownian motion, *J. Phys. A: Math. Gen.* **28**, 15, pp. 4305–4311 (1995), http://stacks.iop.org/0305-4470/28/4305.

[188] G. Shevchenko, Fractional Brownian motion in a nutshell, arXiv:1406.1956 [math.PR] (2014), https://arxiv.org/abs/1406.1956.

[189] I. Calvo and R. Sánchez, The path integral formulation of fractional Brownian motion for the general Hurst exponent, *J. Phys. A: Math. Theor.* **41**, 28, 282002 (2008), http://stacks.iop.org/1751-8121/41/282002.

[190] H. Wio, *Path Integrals For Stochastic Processes*. World Scientific, New Jersey (2013).

[191] S. V. Lawande and K. V. Bhagwat, Feynman propagator for the δ-function potential, *Phys. Lett. A* **131**, 1, pp. 8–10 (1988), http://www.sciencedirect.com/science/article/pii/0375960188906226.

[192] S. M. Blinder, Green's function and propagator for the one-dimensional δ-function potential, *Phys. Rev. A* **37**, 3, pp. 973–976 (1988), https://link.aps.org/doi/10.1103/PhysRevA.37.973.

[193] E. B. Manoukian, Explicit derivation of the propagator for a Dirac delta potential, *J. Phys. A: Math. Theor.* **22**, 1, pp. 67–70 (1989), http://stacks.iop.org/0305-4470/22/i=1/a=013.

[194] J. Mathews and R. L. Walker, *Mathematical Methods of Physics*. Benjamin, New York (1964).

[195] L. D. Landau and E. M. Lifshitz, *Quantum Mechanics: Non-Relativistic Theory, Course of Theoretical Physics*, Vol. 3. Pergamon, Oxford (1977).

[196] L. Elsgolts, *Differential Equations and the Calculus of Variations*. Mir, Moscow (1977).

[197] F. Cooper, A. Khare and U. Sukhatme, Supersymmetry and quantum mechanics, *Phys. Rep.* **251**, 5, pp. 267–385 (1995), http://www.sciencedirect.com/science/article/pii/037015739400080M.

[198] P. A. Binding, P. J. Browne and B. A. Watson, Darboux transformations and the factorization of generalized Sturm–Liouville problems, *Proc. R. Soc. Edinb. Sect. A-Math.* **140**, 1, p. 129 (2010), http://dx.doi.org/10.1017/S0308210508000905.

[199] H. R. Jauslin, Exact propagator and eigenfunctions for multistable models with arbitrarily prescribed n lowest eigenvalues, *J. Phys. A: Math. Gen.* **21**, 10, pp. 2337–2350 (1988), http://stacks.iop.org/0305-4470/21/i=10/a=013.

Index